People Promoting and People Opposing Animal Rights

Recent Titles in
The Greenwood Press "People Making a Difference" Series

People For and Against Gun Control: A Biographical Reference
Marjolin Bijlefeld

People For and Against Restricted or Unrestricted Expression
John B. Harer and Jeanne Harrell

PEOPLE PROMOTING AND PEOPLE OPPOSING ANIMAL RIGHTS

In Their Own Words

John M. Kistler

Foreword by
Bernard Rollin, Ph.D.

The Greenwood Press "People Making a Difference" Series

GREENWOOD PRESS
Westport, Connecticut • London

Library of Congress Cataloging-in-Publication Data

Kistler, John M., 1967–
    People promoting and people opposing animal rights : in their own words / by John
  M. Kistler ; foreword by Bernard Rollin.
        p.   cm.—(The Greenwood Press "People making a difference" series, ISSN 1522–7960)
    ISBN 0–313–31322–9 (alk. paper)
    1. Animal rights.   2. Animal welfare—Moral and ethical aspects.   3. Animal rights activists.
  I. Title. II. Series.
  HV4708 .K48   2002
  179′.3—dc21   2001033682

British Library Cataloguing in Publication Data is available.

Copyright © 2002 by John M. Kistler

All rights reserved. No portion of this book may be
reproduced, by any process or technique, without the
express written consent of the publisher.

Library of Congress Catalog Card Number: 2001033682
ISBN: 0–313–31322–9
ISSN: 1522–7960

First published in 2002

Greenwood Press, 88 Post Road West, Westport, CT 06881
An imprint of Greenwood Publishing Group, Inc.
www.greenwood.com

Printed in the United States of America

The paper used in this book complies with the
Permanent Paper Standard issued by the National
Information Standards Organization (Z39.48–1984).

10 9 8 7 6 5 4 3 2 1

Contents

Series Foreword	ix
Foreword *by Bernard Rollin, Ph.D.*	xi
Acknowledgments	xiii
Introduction	1
Carol J. Adams	13
Ron Arnold	22
David Barbarash	29
Don Barnes	36
Gene Bauston	41
Marc Bekoff	45
Brian Bishop	52
Robert Cohen	62
Priscilla Cohn	72
Karen Coyne	78
Diana Dawne	87
Ryan DeMares	92
Sherrill Durbin	100
Michael Fox	104

Milton M.R. Freeman	112
Margery Glickman	119
Kimber Gorall	126
Alan Herscovici	135
Alex Hershaft	145
J.R. Hyland	152
Roberta Kalechofsky	159
Crystal Kendell	166
Deanna Krantz	172
Finn Lynge	182
Kathleen Marquardt	187
Pat Miller	193
Laura Moretti	199
Ingrid Newkirk	206
Ava Park	212
Teresa Platt	219
Susan Roghair	226
Anthony L. Rose	232
Andrew Rowan	237
Jerry Schill	244
Cindy Schonholtz	253
Mary Zeiss Stange	258
Patti Strand	265
Michael Tobias	269
Frankie L. Trull	279
William L. Wade	283
Ed Walsh	288
Ben White	295

Appendix A: Interview Questions	305
Appendix B: Participant Letter	307
Appendix C: Organizational Addresses	311
Bibliography of Contributors' Writings and Other Resources	319
Index	325
About the Contributors	337

Series Foreword

Many controversial topics are difficult for student researchers to understand fully without examining key people and their positions in the subjects being debated. This series is designed to meet the research needs of high school and college students by providing them with profiles of those who have been at the center of debates on such controversial topics as gun control, capital punishment, and gay and lesbian rights. The personal stories—the reasons behind their arguments—add a human element to the debates not found in other resources focusing on these topics.

Each volume in the series provides profiles of people, chosen for their effective battles in support of or in opposition to one side of a specific controversial issue. The volumes provide an equal number of profiles of those on both sides of the debates. Students are encouraged to read stories from the two opposite sides to develop their critical thinking skills and to draw their own conclusions concerning the specific issues. They will learn about those people who are not afraid to stand up for their cause, no matter what it may be, and no matter what the consequences may be.

To further help the student researcher, the author of each volume has provided an introduction that outlines the history of the issue and the debates surrounding it, as well as explaining the major arguments and concerns of those involved in the debates. The pro and con arguments are clearly defined as are major developments in the movement. Students can use these introductions as a foundation for analyzing the stories of the people who follow.

Greenwood Press's hope is that each student will realize there are no easy answers to the questions these controversial topics raise, and that those on all sides of these debates have legitimate reasons for thinking, feeling, and arguing the way they do. These topics have become controversial because the people involved have very real, emotional stories to tell, and these stories have helped to shape the debates. Each profile provides information such as where and when the person was born, his or her family background, education, what pushed him or her into action, the contributions he or she has made to the movement, and the obstacles he or she has faced from the opposing factions. All this information is meant to help the student user critique the different viewpoints surrounding the issue and to come to a better understanding of the topic through a more personal venue than a typical essay can provide.

Foreword
Bernard Rollin, Ph.D.

Efforts to raise the moral and legal status of animals in society represent one of the most fervid and powerful international social movements of the last three decades. This movement has been remarkably effective in changing accepted practices in virtually all traditional animal uses, from agriculture to zoos. The legislative elimination of confinement agriculture in Sweden; the advent of laws regulating animal research throughout the Western world; the public assault (by referendum) on management of wildlife for hunters; the significant proliferation of animal protection legislation around the world; the rise of university courses on ethics and animals; law school research on raising the status of animals from property; the phenomenal economic growth of cosmetic companies which disavow animal testing; all of these new activities bespeak the power of human moral concern for animals to effect major social change. Yet there has been remarkably little study of this movement, perhaps because it is so often ridiculed and dismissed as "fringe."

What we do know—most animal advocates are female, well-educated, affluent—is based on minute samples and tells us virtually nothing of substance. Above all, interested parties can get no qualitative "feel" for the sort of people who resolutely advocate for animals, or for the inevitable opponents they create.

John Kistler's book helps fill this void by presenting us with qualitative self-portraits, in their own words, of a sample of animal advocates and those dedicated to opposing them. People profiled range from a spokesman for the radical Animal Liberation Front to the long-time chief representative of the biomedical research lobby. All respond to the same questions, and the answers present us with a fascinating tapestry of reasons why people in all strata of society would choose to devote their lives to issues of animal treatment. It is particularly interesting to note how people with similar premises reach wildly different conclusions, as when one feminist is inexorably drawn to veganism, while another sees hunting as the fulfillment of her ideals. In reading some of the statements, it is difficult not to sympathize with Nietzsche's dictum that our metaphysics is a rationalization for our hang-ups. In the end, the book is good reading, both for providing insight into a major social movement and into the minds of those who make it happen.

Acknowledgments

Many thanks to my friends, the Bistygas, for yearly holiday cheer.

Also, I appreciate the many people who participated in these interviews. I am refreshed to see that despite the pragmatism and self-absorption of our modern age, idealism has not been entirely lost.

Introduction

The purpose of this book is to bring together representatives (or activists, as defined below) from many sides of the animal-rights debate, so that students and researchers can see a variety of personalities and points of view in a single printed volume. Dozens of polemical writings are produced each year on animal-rights issues, in journals and in books, presenting their arguments persuasively for one opinion or another. Unfortunately, very few publications have attempted to provide personal perspectives from these activists. This is the first work of its kind, based on collecting interviews with identical questions asked of all participants, so that their responses may be compared and contrasted.

The word "activist" does not imply that these people are necessarily "radicals" or "extremists." Activism has become synonymous with extremism in many media contexts. For the purposes of this volume, *an activist is a person who has taken a larger-than-average role in promoting or opposing a certain issue.* For example, because I spend much time writing books about animal issues, I am something of an activist, though I am not committed to either side. A person who spends a lot of time, energy, and/or money on any given issue could be called an activist. Using this definition, all the participants interviewed in this book are activists, because they strive to influence our society in the cause of animal rights or animal welfare.

My hope is that the razor-sharp edge to many animal-rights debates can be dulled somewhat, if each opposing group can see that "real," well-meaning and intelligent people are working within their belief systems. Perhaps a degree of mutual respect can be attained, which will encourage more cooperation and better reforms. I was surprised, in fact, that the general attitude of the people whom I interviewed was civil and non-hostile. The words "enemy," "evil," or "crazy" do not often appear in this book. Such words are, unfortunately, all that television viewers may see on placards or hear in sound bites at the protests shown on newscasts. Such "loaded" words signify that participants are engaged in a civil war, and are probably not pursuing any meaningful dialogue.

Of great importance to this work is the definition of the terms "animal rights" and "animal welfare." Both phrases have been skewed in meaning by unintentional misuse.

Animal rightists believe that animals are of equal or similar importance to humans, and thus, animals must receive equal or similar treatment to that of humans. In this framework, animals are not to be killed or enslaved by humans, or even used by humans. Using a strict definition like this, even vegetarians are not necessarily animal rightists, since many vegetarians continue to use other animal products (eggs, milk, honey, wool) and keep pets (also called "animal companions"). There is room for debate, of course, as to the strictness of the definition offered above. The Animal Rights Online Newsletter writes, "The philosophy of animal rights maintains that nonhuman animals are 'subjects of a life,' [see *The Case for Animal Rights* by Tom Regan] and as such have an intrinsic right to live free from human exploitation" (A.R.O.N., January 3, 2001). Some animal rightists refer to the keeping of pets as "slavery," while others believe it is a mutually desired and voluntary co-existence between species. However, the animal industries are difficult to defend using any conventional definition of animal rights. Any definition that permits animals to be killed (by experimentation, for food, or for sport) or to be kept in intensive industrial confinement (factory farms) must be, in reality, an animal welfare perspective, not an animal rights perspective.

Animal welfarists believe that humans are of greater importance and value than animals, and that humans can make moral use of animals; but most insist that improvements can (and should) be pursued to improve the animals' lives. Thus, when the title of this book speaks of those "opposed" to animal rights, it in no way means that such opponents are in agreement with intentional cruelty, nor that they oppose all reforms to animal treatment. They simply oppose the philosophy of animal rights that states that nearly all uses of animals are morally wrong. Some animal welfarists might be uncomfortable saying that humans are "superior" to animals, but would in practical matters usually give preferential treatment to the human, if forced to choose between human and animal.

Many people who say that they believe in animal rights are, actually, misinformed animal welfarists. Some animal welfarists declare "I believe in animal rights" because they have misunderstood what animal rights philosophy teaches, and they have not understood what animal welfare philosophy teaches. Without careful distinction between the two groups, it is easy to see how people may become confused about this. For instance, as a boy in the sixth grade, I became very concerned about whaling and carried around a petition from Greenpeace to have other kids sign their names in opposition to harpooning whales. At that time, I saw myself as an animal rights supporter, though in fact I was probably dribbling hamburger grease on the petition at lunchtime. The contradiction had not set in my mind: that I believed whales should not be killed, while I simultaneously consumed a dead cow that someone had killed on my behalf. In truth, then, I was not an animal rights supporter, I was an animal welfarist. I wanted to protect whales more than I wanted to protect cows, which means that I still supported the deaths of many animals for human usage, while simultaneously promoting the salvation of cetacean species.

That is not to say that it is necessarily wrong to save whales while killing cows for food. One could make logical arguments to say that whales deserve special attention because they are endangered, or because they have keener mental abilities and social bonds, or whatever. However, these arguments have already abandoned the key animal rights contention that all animals have equal or similar value to humans, and as such, leaving out cows is not possible. Evidently, many others have made this same mistake, believing that their support of one or two animal rights initiatives makes them animal rightists. Donating money to a local church does not necessarily make me a believer in the philosophies or doctrines of that church; in the same way, donating money to the local humane society does not make me an animal rightist. Membership in a church usually includes a statement of agreement with certain beliefs, and a corresponding lifestyle to match those beliefs. Likewise, to be an animal rightist must include a general agreement with its philosophies of equality of humans with animals, and a corresponding behavior that does not exploit animals.

Admittedly, there are large differences between animal welfare groups. Using the above definitions, all groups which allow the killing or industrial captivity of animals are welfarist groups. This encompasses a vast number of organizations, from humane societies to vivisectionist groups. One way of understanding this is to think of shades of color between black and white: all of that gray! If you view animal rights as white, then animal welfare would encompass the entire spectrum of non-white, from gray through black (black being the farthest possible from animal rights, perhaps groups that support all possible uses of animals). Animal welfare is just the foundational philosophy; there are other large philosophical differences between these welfarist organizations. For instance, perhaps a humane society would oppose animal experimentation and dog racing, but approve of pet keeping and beef or dairy production. Another animal welfare group may support dog racing and experimentation, but oppose whaling and sealing. The variety of groups is mind-boggling. My point is that foundationally, they all have one common denominator, a foundational philosophy, which believes that humankind has the right to use animals for some purposes. This is how welfarists differ from rightists.

Another reason people may misunderstand the philosophies of animal rights versus animal welfare is the nature of media reports on animal issues. These issues are not adequately portrayed in the media—not necessarily out of malicious intent, but due to the complexity of the issues. News organizations try to simplify issues, and it is difficult to explain the levels of gray amid the blacks and whites. Many animal industry producers do care about the animals under their control. However, when you have one minute of television time to show both sides, what will be shown? Perhaps thirty seconds of protestors saying "meat is murder," and thirty seconds of a farmer saying "they're entitled to their opinion, but people need to eat." How many of the important details of the real issues are missed in such a report? Frankly, almost all of them. How many of the protesters were wearing leather items, and how would they defend this practice? Do they eat eggs or drink milk? Or, how does the farmer justify the injections of hormones and antibiotics into his cattle? Is the animals' confinement excessively small? There are myriad issues to be considered; these are complicated matters. Sound bite answers only muddy the waters; they do not clarify anything, they simply give the illusion of providing answers. In this situation, the farmer is almost automatically on the defensive: he/she is being asked to prove why his/her activity is proper. In such an environment, it is no wonder that people who feel compassion for animals would want to be called animal rightists; because animal rightists are often portrayed as the people who care about animals, while animal welfarists are purported to be the people who do not care.

The key misunderstanding among people who wrongly believe themselves to be animal rightists is this: they think that because they desire improvements in animal lives, and reductions in animal pain and suffering, they must believe in animal rights. This is an incorrect assumption. *Animal rights philosophy promotes the removal of animals from captivity and from industries. Animal welfare philosophy promotes the improvement of animal life in captivity and industries.*

Ironically, animal rights groups use anti-cruelty campaigns to gain financial support from confused animal welfarists; while animal welfare organizations are wrongly labeled "pro-cruelty" (when they are noticed at all). Such controversy is not limited to the animal rightists' camp. Animal welfarists are also surrounded by controversies.

First of all, viewers can be suspicious when they find that some animal welfare groups have their chief means of financial support from funds of the major animal industries: beef, dairy, poultry, entertainment, and other subsidiaries thereof. That is not to say that the companies are necessarily dishonest, and use the groups only for good public relations, but it is easy to suspect ulterior motives; just as the public is suspicious of anti-smoking advertisements that are paid for by the tobacco industry. Secondly, the animal welfare industry and its supporters can be very suspicious of the media, publishers, and researchers, thus denying themselves valuable opportunities to spread their own viewpoints. No doubt, they have been deceived by unscrupulous or biased sources. However, by reacting defensively, and not participating in books or interviews, they 'look guilty,' like defendants who say "no comment" when asked questions. Perhaps my own attempts to contact certain animal industries were flawed, but in several cases I received rather hostile rebuffs. When I offered clarifications and offered a revised form to meet some of their concerns, only one reconsidered and joined this project; the rest continued to strongly oppose any participation. One industry group insisted that it must create and answer its own questions with no editing permitted. I am grateful to those who did participate and do not mean to demean their efforts; I am simply pointing out that animal welfare groups do not generally make good use of the media.

For the curious among you, I am undecided on these issues. Of course I have opinions and lean a certain direction on several animal issues; however I am not decided upon the large question, do I believe in animal welfare or animal rights? I am definitely, at least, an animal welfarist. I am a guilty meat eater, meaning I still eat meat, but only once a day, and I admit to being unsure about the virtue of that meal.

On the other hand, I am fairly certain that humans need to use animals for some tasks, even if we do not kill them. So at the moment I am an animal welfarist, but I am not certain that animal rights philosophy is wrong. I am still pondering these issues. Thus, I would hope that my work does not seem biased toward one side or another. Of course, as a scholar and librarian, it would be unethical for me to intentionally skew evidence to one side, even if I were a "true believer" in one side.

To choose the interview questions I used for this volume, I wrote a long list of several dozen possible questions that would gather information about participants' feelings, motivations, goals, thoughts, histories, and philosophies. Next, I grouped these possible questions into larger categories; eliminated duplicates; combined questions that were similar; and condensed the list to include one dozen queries. Greenwood Press editors discussed these, and suggested several changes. I made improvements based upon their comments.

In October 1999 I sent out these interview questions to about two dozen candidates as the initial test run. With the questions, I also included an explanatory letter about the project and its goals. The animal rights supporters were generally quick to accept a role in the book, but animal welfarists were often reticent or even hostile. At first I felt a bit defensive myself about being "scorned," but eventually realized that they have probably been abused by the media. Could my questions have unconsciously been biased toward the animal rights side? I doubted it because I am a moderate, at best. Furthermore, it was too late to change my questions because my editors had approved them, and several people had already replied with their answers. So I came up with an alternate idea, being unable to change the questions. I offered everyone the option to challenge my questions, with their own alternative questions, in the text of their own responses. In other words, my questions had to stay, but any participant can say in their own answers, "that is a bad question, the real question is. . . ." This addition to my query letters brought about an increase in participation among animal welfarists. In reality, not many people used this option, but having the option may have softened the fear that my intention was to trap them with clever editing or tricky questions. You can read the query letter that was sent to each invited participant at Appendix A.

I knew that there would be some suspicion on all sides, especially since these questions are more personal, and not simply opinion pieces. I included my credentials, the name and address of my Greenwood

Press editor, and the names of some other participants in the interviews. This was especially important to animal welfarists, who would rightly fear being the lone scapegoats in a book if it was to be dominated by animal rightists! Fortunately I was able to come up with an even split of animal rightists and animal welfarists for the interviews. In our society, animal welfarists must outnumber animal rightists by ten to one, at least. But animal welfare organizations are not as well organized as other special-interest groups, and most do not wish to be a part of the public debate. This fear of public exposure by animal welfarists has become more acute in recent years, because of fear of retribution by "direct action" groups like the Animal Liberation Front. In truth, violent attacks upon humans by animal rights "extremists" are very rare, and except for a few anonymous essays, such attacks are not promoted officially by any animal rights group. On the other hand, having protestors demonstrate in front of your home or workplace could be an unpleasant experience, and thus it seems wise to many animal welfarists to remain out of the public eye.

Another means of allaying suspicion among my potential participants was to assure them that my role as editor would be rather minimal. I chose participants, collected their interviews, and then checked their responses for proper grammar, spelling, readability, relevance, and space. I sent the edited piece back, so that the participants could see any changes made, and made any further desired changes with their permission. I have no interest in altering meaning, or choosing only the catchiest, most controversial statements, as a "shock-jock" might do for a television talk show. Furthermore, Greenwood Press requires signed permissions of approval from each participant, agreeing that the final draft of the interview I sent them was agreeable and accurate to their work. It would be unlawful (and unethical) for a publisher or an editor to misrepresent a participant's interview.

The purpose of this book is to show what activists are like personally. We already know, to some degree, what they believe. Why do they think this way? What events led them to this conclusion? What goals do they have? How do they balance their activism with their "normal life?" These are questions that have not been adequately answered, in my opinion, in earlier works. I will go over the questions and my reasoning in choosing each one here in the Introduction. To reduce redundancy (the twelve questions repeated forty times in these pages), one of the Greenwood editors suggested using a condensed form of the question. So, the reader will recognize "Q.1 Biographical Profile" as referring to Question One, which is the autobiographical question.

INTERVIEW QUESTIONS

Q1. Could you tell us a little bit about yourself? Who are you? Where and when were you born? What do you enjoy? [BIOGRAPHICAL PROFILE]

The answer to this question simply establishes a brief background and history of the interviewed individual. His or her approximate age, interests, and hobbies may be of interest to researchers and the layman. Activists often have activities that are not centered wholly on animal issues.

Q2. How did you become involved in animal rights issues? Was it a single event, or a gradual process that started you down the path toward activism? [BECOMING INVOLVED]

The human mind often desires to find a cause-and-effect relationship to explain behavior, and often there are some "triggers" that motivate a person to become active in an issue. In some cases it is the witnessing of a single wonderful (or horrific) event that spurs the individual to become involved. In other cases, it is a long process, where a chain of happenings leads the person to actively seek changes. Neither case is "better" than another; our hearts are stirred in different ways.

Q3. What are one or two issues that you spend the most time on? Why do you care so much about this (or these) issues? Why should other people also care about this? [IMPORTANT ISSUES]

I tried to find participants from many different areas of animal-rights issues. Of course there are people who care about more than one issue; but in general, activists have found a "niche" area that they choose to focus their energies on. There are so many animal issues that not every type could be represented in this book; however, these interviews do cover a number of the key debates we face in our culture.

Q4. What are your short-term and long-term strategies for achieving your goals? What do you do exactly, on a regular basis, toward these goals? Write? Research? Educate? Protest? [GOALS & STRATEGIES]

Most activists have some strategies for achieving their goals, by using their talents in a directed manner. Many choose to engage in public protests; others write articles and books; and others do research to find hard facts that will support the work of others who will publicize it.

Q5. What is your ultimate goal on this issue? When could you say, "we have succeeded"? [DEFINING SUCCESS]

The ultimate goal is the far-distant (possibly unreachable) ideal that inspires an individual's efforts. Most activists have a reason for their work—the problem that drives them to action. This ideal, desirable future is their ultimate goal. In some sense, this is the boiled-down, core motivation for the activist. Aside from all the tangential issues that deal in the specifics and practicalities, the ultimate goal is the center toward which all of their efforts are directed.

Q6. Do you work closely with any formal groups or organizations? Which one(s)? [GROUPS & ORGANIZATIONS]

The participants can identify organizations that promote their own strategies and ideals.

Q7. What groups or types of people do you consider to be "the opposition?" ["THE OPPOSITION"]

The participants can identify organizations or individuals or belief systems that they view as the cause of the problems that they wish to solve. Many participants do not like the label "opposition" and choose friendlier terms; but all activists, by definition, have goals, and therefore have belief systems and/or persons who are opposed to those goals.

Q8. Whom do you admire the most, or who inspires you? Name one or two, modern or ancient heroes you have. [HEROES]

The individual's examples of people he/she admires may inspire the readers to further research and emulation. These examples may also show what characteristics, methods, and beliefs that the participants themselves see as laudable or worthy of imitation.

Q9. Is religion or spirituality a part of your life? Does this religion (or lack of it) help, motivate, or hinder your work? [RELIGION/SPIRITUALITY]

Many studies of activists seem to ignore religion as a key factor, though many or most activists claim to be religious or spiritual. Researchers tend to avoid discussion of religion because they view it as subjective and non-verifiable, but in doing so they ignore an important part of the activist's motivation. Religion is an organized and formal method of living out a cherished philosophy or worldview. Everyone has a philosophy of life, even an anarchist. Similarly, everyone has a spirituality of some sort—even atheism is a life philosophy. The interesting question in this context of animal rights issues is: did these participants choose a religion based upon their views on animals, or did their early adoption of a religion lead to their current views on animals? Which came first, the foundational philosophy or the specific opinions? Perhaps it does not matter; in either case the individuals believe that they are now doing the right thing, or the consistent thing, within their philosophies.

Q10. Would you describe a very funny or very strange experience that you have had in your work as an activist? [FUNNY/STRANGE EXPERIENCES]

The purpose of this question is to lighten up the often-depressing stories and statistics that crop up in most animal-rights discussions. Of course, these are serious issues, and not a laughing matter. However, human beings cannot function well without humor, and I believe that the over-seriousness of some activists may actually be detrimental to their persuasiveness (not to mention their own psychological well-being). After years of working in a certain field, strange or funny things crop up from time to time, and this question provides an opportunity for the participants to share their unusual encounters with readers.

Q11. What advice would you give to a person just starting into animal rights (or opposing-animal-rights) activism? What important lessons have you learned; or mistakes have you made, that others might learn from? [LESSONS/ ADVICE]

One fringe benefit that readers may gain from this book is the free distribution of hard-earned advice from experienced activists or workers. When participants share their victories and defeats with us, we may be able to improve upon our own efforts with less "trial and error" and more success.

Q12. Do you have anything to add to these questions that might be helpful to readers? [ADDITIONAL INFORMATION]

Just in case the prior eleven questions failed to address some important aspect of activism, question twelve allows the participants to touch upon other issues that were not specifically addressed.

How were participants selected? Doing research for my first book, *Animal Rights: A Subject Guide, Bibliography, and Internet Companion* (Greenwood Press, 2000), I learned much about the issues, groups, and leaders of each viewpoint. Whenever possible, I contacted leaders or well-known spokesperson from each corner of the debate. Once I found a spokespeople from one point of view, I tried to find a person to represent the opposing viewpoint. In many cases, participants made helpful suggestions and even recommendations to help me find such opponents, leading me to contacts I might not otherwise have made. I sought to find a good mixture of scholars and front-line workers: so that the book would have a variety of personality types represented. This has the further advantage of diversity in the styles of writing and life experiences that each participant can share with us.

Carol J. Adams

Q1. BIOGRAPHICAL PROFILE

I am an activist, a writer, a vegan, a feminist, a partner, a parent, and a yoga practitioner. I grew up in a small town in upstate New York during the 1950s and 1960s. I have been involved in issues of social justice since the 1970s. After receiving a Master of Divinity from Yale University Divinity School in 1976, my partner and I started a Hotline for Battered Women in Chautauqua County, New York. For the first eighteen months, it was housed in our home. During that time, I was the executive director of the Chautauqua County Rural Ministry, Inc., an advocacy and service not-for-profit agency addressing issues of poverty, racism, and sexism. During the next decade, I served as chairperson of the Housing Committee of the New York Governor's Commission on Domestic Violence (1984–1987), coordinated two legal challenges (against a local radio station and against a city) for racism, and I began writing *The Sexual Politics of Meat*.

Since 1987, I have lived in the Dallas area. I developed a course on "Sexual and Domestic Violence: Theological and Pastoral Concerns," which I have taught at Perkins School of Theology, Southern Methodist University. I also authored a book about woman-battering to help ministers. I am particularly interested in the interconnections between forms of violence against human and nonhuman animals, and have written about why woman-batterers harm animals and the implications of their violence. Recently I received awards from The Greater Dallas Coalition for Reproductive Freedom and Planned Parenthood of Dallas and North Texas, "for her help in understanding the psychology of the

Carol J. Adams. Photo credit: Kate Sartor Hilburn

radical right, for her commitment to women and for her brave stance against the tyranny of Operation Rescue."

I enjoy discovering and expressing connections—connections, for instance, between feminism and vegetarianism (as explored in *The Sexual Politics of Meat: A Feminist-Vegetarian Critical Theory*), and between vegetarianism and spirituality (as explored in my most recent book, *The Inner Art of Vegetarianism: Spiritual Practices for Body and Soul*).

I love cooking and sharing vegan meals, knowing that no animals were harmed for the meal; in this way, cooking and sharing food can truly embody compassion. I need walks in the woods, and living in Texas now, I miss the smell and sound of crinkly maple leaves on the floor of an upstate New York forest.

Q2. BECOMING INVOLVED

At the end of my first year at Yale Divinity School, I returned home to the small upstate town where I had grown up. As I was unpacking,

I heard a furious knocking at the door. An agitated neighbor greeted me as I opened the door. "Someone has just shot your horse!" he exclaimed. Thus began my political and spiritual journey toward a feminist-vegetarian critical theory. It did not require that I travel outside this small village of my childhood—though I have; it involved running up to the back pasture behind our barn, and encountering the dead body of a pony I had loved. Those barefoot steps through the thorns and manure of an old apple orchard took me face to face with death. That evening, still distraught about my pony's death, I bit into a hamburger and stopped in mid-bite. I was thinking about one dead animal yet eating another dead animal. What was the difference between this dead cow and the dead pony I would be burying the next day? I could summon no ethical defense for a favoritism that would exclude the cow from my concern because I had not known her. I became aware that I was a meat eater; simultaneously I realized that with this awareness that I was eating animals, I needed to stop eating animals.

After becoming a vegetarian, I thought to myself: Why am I concerned about animals dying to be food and not animals experimented upon? Once I became involved in animal activism, I asked myself why I was concerned about animals and not the rest of nature. And so I became involved in ecofeminism, a movement that links concerns with the natural world with feminist concerns for justice, nonviolence, and the honoring of aspects of life that are traditionally female-identified.

Q3. IMPORTANT ISSUES

I spend the majority of my time on the issue of vegetarianism. I care about this issue because it contains so many of the issues that now trouble our planet. A meat-eating culture places tremendous environmental demands on the natural world (in terms of use of oil, deforestation, pollution of water, destruction of the ozone layer). I wish to see social justice and a more compassionate world, and I know that in order for meat eating to exist, the poor, people of color, and women are hired to work in slaughterhouses and do the dirty work of killing and packaging animals. I am concerned about our health (six of the ten leading diseases that kill Americans are directly linked to meat and dairy consumption). And I have a deep intuition that we are meant to do the least harm possible. We can live more healthily on a plant-based diet, so why should we require the suffering and deaths of billions of animals?

Because I am a person who thinks philosophically and theoretically, one of my contributions is to provide a theoretical structure that explains people's resistance to the message of vegetarianism and animal advocacy. I do this through my books. *The Sexual Politics of Meat* (10th anniversary edition, Continuum, 2000) argues that our view of animals is largely built upon the way that gender politics is structured into our world. This is especially true for our view of the animals that are consumed. Patriarchy is a gender system that is implicit in human/animal relationships. Moreover, gender construction includes instruction about appropriate foods. Being a man in our culture is tied to identities that they either claim or disown—what "real" men do and don't do. "Real" men don't eat quiche. It's not only an issue of privilege, it's an issue of symbolism. Manhood is constructed in our culture, in part, by access to meat eating and the control of other bodies.

Everyone is affected by the sexual politics of meat. We may dine at a restaurant in Chicago and encounter this menu item: "Double D Cup Breast of Turkey. This sandwich is so BIG." Through the sexual politics of meat, consuming images such as this provide a way for our culture to talk openly about and joke about the objectification of women without having to acknowledge that this is what is being done. The sexual politics of meat also works at another level: the ongoing superstition that meat gives strength and that men need meat. There has been a resurgence of "beef madness" in which meat is associated with masculinity.

Meat eaters like to believe that they are doing what vegans do—eating humanely—without actually doing what vegans do—not eating animal products. My book, *Living among Meat Eaters* (Three Rivers Press, 2001) proposes that meat eaters are blocked vegetarians. In other words, we should view meat eaters as wanting to be vegetarians, but something is holding them back. Their reactions to vegetarianism show us the things that are causing them to be blocked.

I believe that meat eaters would prefer to eat more healthily without requiring the suffering and death of animals, but they think they are unable to change. I see part of my work as helping meat eaters relinquish their need to eat meat by showing them how gentle this change can be. That is the message of my books that form the series of "The Inner Art of Vegetarianism: Spiritual Practices for Body and Soul." The first book in that series explains the deep interconnections between vegetarianism and spiritual practice. In the companion workbook, I provide exercises, meditations, and writing prompts to help someone move along a path toward a plant-based diet. Finally, *Meditations on the Inner Art of Vegetarianism* affirms each person's spiritual journey toward wholeness.

Q4. GOALS & STRATEGIES

My strategies are to write, to teach, to interpret, to patiently listen, and to provide people with the internal space to change.

I travel to universities and colleges and show the "Sexual Politics of Meat" slide show, I write articles and short commentaries for a variety of publications, I speak to vegetarian groups, and I write a column on "Living among Meat Eaters" for *The Animals' Agenda* magazine.

Q5. DEFINING SUCCESS

My goal is to live in a world where meat eating is as rare as cannibalism. However, because of the role of imperialism and its impact on native cultures, I would be happy when the developed countries, specifically the United States and Western Europe, had converted to a vegetarian diet, and the U.S. government stopped providing federal support for meat and dairy production.

When will I be able to say "we have succeeded?" When people discover for themselves the life-changing and life-affirming qualities of vegetarianism.

Q6. GROUPS & ORGANIZATIONS

I work with a variety of animal activist groups depending on the issue, but most specifically, with Feminists for Animal Rights (FAR). FAR seeks to raise the consciousness of the feminist community, the animal rights community, and the general public regarding the connections between the objectification, exploitation, and abuse of both women and animals in a patriarchal society. As they state, "as ecofeminists, we are concerned about cultural and racial injustice and devaluation and destruction of nature and the earth. We view patriarchy as a system of hierarchical domination, a system that works for the powerful against the powerless. FAR promotes vegetarianism and is vegan in orientation. FAR is dedicated to abolishing all forms of abuse against women and animals." They publish an *Ecofeminist Journal* that explores the interconnections of activism among feminism, environmentalism, and animal rights. They offer a "FAR Marketplace," where feminist-animal rights resources are available for purchase. And they have pioneered a foster care program for the companion animals of battered women, so that the women may enter a battered women's shelter and yet know that her batterer cannot harm the animals she cares about.

Q7. "THE OPPOSITION"

At the primary level, the opposition is the companies that use animals to produce food: meat, dairy and egg companies. At the secondary level are all people who support those companies by eating meat, dairy, and eggs, rather than understanding that a boycott of these companies is necessary. Meat eaters will buttonhole me demanding, "What about the homeless, what about battered women?" and insist that we have to help suffering humans first. I always find this ironic: I know that this question is actually a defensive response, an attempt to re-establish the meat eater on higher moral ground than the vegetarian. In fact, only meat eaters raise this issue. No homeless advocate who is a vegetarian nor battered-women's advocate who is a vegetarian would ever doubt that these issues can be approached in tandem. In addition, the point of my work, basically, is that we have to stop fragmenting activism; we cannot polarize human and animal suffering because they are interrelated.

Q8. HEROES

I admire Susan B. Anthony and Elizabeth Cady Stanton, nineteenth-century social activists, who worked for women's rights. Anthony was a strategist; Stanton, a wonderful philosopher. They understood two important things—when one is reforming a basic, accepted aspect of cultural life, it takes a long time. One must be ready to be in for the long haul. Second, they recognized the importance of having the courage of one's convictions. "Cautious, careful people will never bring about a reform," Anthony observed. After working on the issue for more than fifty years, they both died before women had gained the right to vote. Persistence and vision are their legacies.

I also have two contemporary heroes: Marie Fortune, the founder of the Center for the Prevention of Sexual and Domestic Violence, who for twenty-five years has pioneered in educating religious communities about sexual and domestic violence; and my partner, Bruce A. Buchanan, whose ministry to those who are usually left out (especially the homeless), teaches me daily the importance of safety, patience, and perseverance, not to mention the fact that effective change happens one situation, one person, and one issue at a time.

Finally, all the women who have worked in animal advocacy inspire me. I remember a comment by one animal activist who felt that now the animal rights movement had demonstrated that "We're no longer just little old ladies in tennis shoes." For myself, I am hoping to live long enough to be a little old lady! Perhaps, when those women started working on animal issues, they weren't little old ladies either! Even today, an estimated 75 percent of activists are women. It would be nice not to be a little old lady in tennis shoes working on the issue of respecting and honoring animals. But that really isn't up to me. It is up to the people who harm animals. My challenge is to find ways to teach, educate, welcome, and invite people into experiencing the wholeness that comes from deciding to stop being an oppressor.

Q9. RELIGION/SPIRITUALITY

In *The Inner Art of Vegetarianism*, I make the case for the deep spirituality of vegetarianism and propose specifically a conceptual framework I call "Spiritual Vegetarianism." Spiritual vegetarianism arises from a desire for wholeness; it is a spiritual practice that links us to the rest of nature and the rest of our own nature, that acknowledges the interconnectedness of all beings and enacts compassion toward them, and that is a living Ahimsa, the absence of violence.

I believe that relationships with animals touch a very deep place within us. When we close off our relationship with that very deep place (either because it is too painful or because we don't give it the time), we are closing off the possibility of wholeness for ourselves. If we eat meat, we may close off our relationship with that deep place, and may not be able to go near it, because doing so would mean we would have to become aware of what we do to animals.

Religions can become a roadblock to vegetarianism. People wrongly interpret the dominion over animals of Genesis 1:26 to justify every cruel act they inflict on animals. Some of my writings specifically, *Ecofeminism and the Sacred*, address this issue.

Q10. FUNNY/STRANGE EXPERIENCES

The reviews to my book, *The Sexual Politics of Meat*, were a big surprise. I have amassed a file of several hundred reviews of the book. Many disagreed vehemently with it—but they spent lengthy paragraphs doing so! The most enjoyable example of this was a long review by the British essayist and critic Auberon Waugh in the *Sunday Telegraph*. He speculated that a male academic émigré from Eastern Europe, who poses as a madwoman (me), conceived the entire book, the author, and her family! I had a good laugh when critics complained that *The Sexual Politics of Meat* proved that the political left still did not have a sense of humor. What they meant is that I did not have their kind of humor. I learned that people did not agree with me, but that they took the ideas seriously enough to disagree with the book. That in itself was an accomplishment!

Q11. LESSONS/ADVICE

This is a long process and requires patience. We are not going to see change overnight. This means that you must take care of yourself as you promote change. We need activists who understand that compassion begins with ourselves, too. Because we're all related, our compassion for ourselves creates the ability to feel compassion for the suffering of others and the desire to relieve suffering rather than inflict harm.

We should ask questions of our activism:

Does it cause fear? Actions that cause fear will be self-defeating, because fear triggers something very deep in terms of our bodies, and this closes off access to the person. When the physical body is coping with stress, fear causes people to feel defensive, and they feel they must stand their ground.

Does it objectify individuals? Does it fail to acknowledge their feelings or their life? One way to show people animals' lives is through animal sanctuaries. Animal sanctuaries allow people a safe environment in which to experience their relationships with animals. Simultaneously, sanctuaries witness against what is being done to animals. When people experience troubling information in a safe space, they are more able to receive it.

Does it use humor? Campaigns that use humor can often bypass conscious defense mechanisms and provide a moment of intuitive connection. "Aha!"

Does it raise consciousness through perception, rather than through words themselves? Like humor, visual images reach us and teach us in ways that words alone can't. This is a potent form of education. Its strength is obvious in the fact that many images that depict animals' suffering have not been allowed on billboards or in other forms of advertisement. This indicates how powerful a vision of the suffering animal can be. For instance, United Poultry Concern's Karen Davis uses a replica of a battery hen cage with five artificial but very realistic-looking hens crammed into it. Each of the hens has a facial expression. People approach the cage, wondering: "Why are these chickens in there?" Perception leads the way, and then they hear the answer: "This is how 95 percent of the chickens who lay eggs live." Attention leads to caring, because it is clear that five hens in a battery cage causes much suffering. The individual encountering his own perceptions realizes: "Now that I've seen one bird and seeing made me understand, I do not have to see every bird to know that this should be challenged." The conceptual flows from the perceptual.

Does it acknowledge commonalities or shared intentions? Many of activist Henry Spira's tactics did this, bringing people with disparate interests together to focus on one common goal. Spira, therefore, allowed everybody to feel that they had an investment in winning and, more importantly, that there would be no losers.

Q12. ADDITIONAL INFORMATION

I am probably best known for introducing the concept of the "absent referent." Behind every meal of meat is an absence: the death of the animal whose place the meat takes. This is the absent referent. It is that which separates the meat eater from the animal and the animal from the end product. The function of the absent referent is to keep our "meat" separated from any idea that she or he was once an animal, to keep something from being seen as having been someone. People are much happier eating something than someone. Many of my books arise from this insight about the absent referent and apply this insight to the various cultural, emotional, and spiritual situations that arise because of the existence of the absent referent.

You can find more information on my Web site at *www.caroljadams.com*.

Ron Arnold

Q1. BIOGRAPHICAL PROFILE

I am Ronald Henri Arnold, born August 8, 1937 in Houston, Texas. My grandparents raised me.

After one year of college, I wandered around the United States, not enjoying school much. I worked for Boeing (in Seattle) for ten years, and started my own business as a consultant. In 1984 I signed on as executive director for the Center for the Defense of Free Enterprise, a 501C3 educational organization in Bellevue, Washington. I serve without compensation, which is why I keep my consulting firm.

I have been married for twenty-five years, and have several children and grandchildren.

Ron Arnold.

I love music, especially classical. I love books and have many thousands of them. As a writer, I refer to many books as sources for finding important information. Books and music are my long-standing hobbies.

I enjoy traveling. My wife and I often go to the opera.

Q2. BECOMING INVOLVED

I got involved with animal rights issues under the old maxim, "Life is what happens while you were making other plans." I had no interest in it originally. But as an executive director for this business organization, I have members, and must pay attention to the issues they face. About ten years ago, in 1989 or 1990, I began getting calls. One was from ranchers in New Mexico, "Somebody just shot my cows," or from a TV repair shop in Massachusetts, saying that protests against surgical supplies have clogged the streets and are harming business, because people couldn't get into his shop. After several of those, it became clear to me that I was not seeing a random assortment of things, but that there was a concerted action on animal rights issues. I began reading up on the standard materials, to see what this movement was.

There is some debate as to whether the animal rights movement is different than the environmental movement. These are complex movements and cannot be easily characterized. There are many strands, and the unlawful part, finding perpetrators, is difficult. There are buckets of civil disobedience type acts, misdemeanors, blocking buildings. Then there are the more violent ones, such as Rodney Coronado. I have read all the court documents. He was with the Animal Liberation Front, and also reported to Ingrid Newkirk, working to find the Silver Springs monkeys. False identifications were found.

In my opinion, the famous book *A Declaration of War: Killing People to Save Animals and the Environment* (written under the pseudonym "Screaming Wolf") was something of a decoy. The two people who wrote and published it (though they claim to have received it anonymously in the mail) in 1991 then fled to Canada. The book is illegal in Canada, as inciting to murder, but it is legal in the United States. It has no importance; the writers understood that for the mainstream movement to succeed, it was essential to have (or pretend to have) people further to the extreme. It was simply trying to plant an idea. Like in Congress, a congressman may offer a bill, knowing that it has no hope of passing in the near future, but it is to put the idea on the table. That book was just to put the idea on the table. They may have hoped that some person would take the book's advice and run with it. The title is really all you need. If you analyze this movement over many years, you look less at the content and more at the context: Who is doing it, why, who benefits, what really resulted from it? Have we seen any murders from this book? No. But some violent acts are occurring. Recently a pair of men planted five pipe bombs under mink transporting trucks. We are fortunate that no one I know of has actually been murdered.

Q3. IMPORTANT ISSUES

Most of the time I am responding to criminal acts. The reason I care is that they are hurting my members. The Center for the Defense of Free Enterprise is a nonprofit organization with about 10,000 to 15,000 members in the United States. The majority of members live in rural areas, which is also where many animal rights attacks are based.

We provide information, we sue people for crimes. We have the EcoTerror Response Network. When people get anonymous tips on planned attacks, we report it to law enforcement authorities.

Q4. GOALS & STRATEGIES

We lobby Congress to make penalties for animal enterprise crimes more severe. Right now the demarcation is $10,000 damage. We would like it to be for any amount of damage, though realistically we may get it down to $5,000 damage for the demarcation line. We are working with Senator Orrin Hatch on the Judiciary Committee, who appreciated my book *Ecoterror*. We want the penalties to be more serious. Essentially the Clinton administration has given orders to law enforcement agencies not to enforce laws against animal rights people. There was a congressional investigation about Warner Creek, a logging area that was picketed and seized by environmental activists (for some endangered species, probably an owl). The Earth First–type folks from Eugene, Oregon, took over the site (despite legal logging permits). The Clinton administration ordered that no actions be taken. The roads were spiked and trenched. That report is available through the House Resources Committee.

Q5. DEFINING SUCCESS

My ultimate goal is that business as usual can proceed without interference from people who do not like animal enterprises. Now realistically that will never happen, because so far this is a country that does allow dissent, and the First Amendment has not been repealed (nor would we wish it to be). As long as there are violent vegetarians, there will always be a threat against these animal enterprises. There will always be people wanting to eat animals, and violent vegetarians wanting to stop people from eating them. Going back in history, even in the earliest human civilizations, activist vegetarianism was to be found. Even the *Epic of Gilgamesh*, which may be the oldest of written human traditions, contains a similar idea. The very first part is the story of Enkidu, a wild person who lived among the wild animals. According to many of these stories, Enkidu tears up the human traps to keep them from catching animals.

Q6. GROUPS & ORGANIZATIONS

I am the executive vice president of the Center for the Defense of Free Enterprise (the president is Alan Gottlieb). Aside from my work with that organization, I am not engaged in animal rights issues except as an author.

My book *Ecoterror* starts with analysis on the Unabomber, then proceeds to other violent activism, including environmental and animal rights efforts. Terrorism is difficult to prosecute. Although we recognize it, it is not actually a crime in the United States, because you cannot criminalize social protests . . . actually law enforcement simply prosecutes the common crimes involved, such as trespassing, possession of weapons or explosives, robbery, and so on. Most of these acts start with robbing banks (for funds). Not until they are found guilty of the smaller crimes can a judge consider the motives, and perhaps this has an influence on the sentencing. Terrorism laws are really in their infancy and unclear in the United States. The FBI generally looks at any crime committed by one person (regardless of his/her motives) as non-terroristic, that is, serial killing. Only when a group is directly involved is an action considered possibly to be terroristic. I studied about twelve hundred crimes actually committed (six hundred by animal rights people). The earliest animal rights crimes I can find started in the 1930s.

Q7. "THE OPPOSITION"

Realize that you are really fighting for something, not against something. Animal rights people come from a viewpoint that could be best labeled "primitivism," thinking that an earlier or simpler way of living is better. They want to force that simpler way, not harming animals in any way. You are really a proponent of the project of modernity. That is a phrase of Habermas, whom I admire. The project of modernity is not just industrial civilization but also the mind styles that go with it: the "let's try it and see if it works," rather than the "don't make any mistakes" mind-set. Those are totally different mind styles. You see a lot of people who want to take the rule of prudence (avoid errors), but the only way to do that is to die. Whatever agenda they have to shut down this or that industry, they use this rule of prudence. There is no such thing; it is nonsense. Realize that you are not anti–animal rights, but for the project of modernity and industrial civilization. Let's complete that project, let's find ways to do more with less, let's find ways to have more people and less impact on the world. Let's find ways to wisely use the earth. Like it or not, human beings are the dominant organism on this planet. You may think this is good or bad. Some compare humans with a cancer or a plague; that is a very unhappy vision! Talk about low self-esteem; that must be as low as it gets.

Q8. HEROES

The older I get, the fewer heroes I have, because the more you study a person the more you come to recognize their flaws. They are as much victims of human frailty as we are.

One of my heroes was Thomas Jefferson, a defender of private property and other such things. But his statements were not always consistent with his actions.

I admired David Brower, who was a personal friend. I agreed with very few of his ideas. He was executive director of the Sierra Club for many years, and he founded Friends of the Earth, then Earth Island Institute. We crossed swords on TV at times, but then went to the bar and enjoyed our odd relationship. We had respect and admiration for each other.

I also admire Aristotle who brought rationality to western civilization.

Q9. RELIGION/SPIRITUALITY

My grandparents, who raised me, had very contrary views on religion. One of them was quite the atheist, but the other made me go to church every week. I gained an intellectual grasp of religions (studying them thoroughly), but never really understood them. But as far as spirituality goes, a sense of awe and wonder, these have always been a part of my life.

I am full of awe and wonder for all of existence. Are all these religions true? Yes, they probably are, but I have never been part of them. Many people take offense on religious grounds; this is a great stumbling block for those who oppose animal rights because it blinds them. They look inward instead of at the outward evidence or arguments. They feel a personal affront, which I do not feel. So having a religious background is probably not much help with those who have no such religious feelings. I know that many animal rights people are religious. Some have become gloomy, and dead serious, without joy or connection to other people. This is not desirable.

In order to grasp the animal rights movement, you must grasp a lot of their ideas on spirituality. I advise reading *Drawing Down the Moon* by Margot Adler, which explains a lot of things using strong field research.

Know thyself, and know thy enemy.

No, I do not have any religiosity, but I do have wonder, which is spiritual, and I think that there is good evidence for nonmaterial forces in the universe. What are they? I am not sure.

Q10. FUNNY/STRANGE EXPERIENCES

I wrote a joke book about this called *Politically Correct Environment*, so if you want lots of jokes on animal rights, you can read that book.

We do have a sense of humor about these things. If you take it too seriously you could go nuts. But if you look at it from the victim's side, when you find out that your $12 million ski resort got burned down to save the lynx, you have no sense of humor about it. Fortunately no one has been killed.

Q11. LESSONS/ADVICE

Don't give in to hate. It is easy to do this. Animal rights people are by and large consumed by hatred of those who do things to animals that they do not like. Most people have a highly developed love for animals and do not want to hurt them. But they still want to eat McDonald's hamburgers. So they keep cows penned up for the cheese, and cows die so they can eat the muscles. These are good-hearted people. No guilt.

Animals do not have rights. Rights are a human construct, not an animal construct. Animals have only as many rights as humans want to give them. Rights are a judicial notion. If you remove humans from the universe, there would be no such thing as rights. There would only be animals that kill each other to get energy. Notions of this kind of morality without humans just do not exist. The idea that animal rights exist is an illusion.

Humane treatment? Sure, you can make a law, and you have given the animals some right judicially.

Be true to yourself. If you want to eat meat, do it. If you don't want to use animals, don't. Either way, be true to yourself. But if you think you have a right to demand that other people conform to your wishes, you have stepped over the line. It is a conflict of values that cannot be resolved. I don't care what they believe anymore than they care what I believe. They can behave the way they wish, but they cannot tell me how to act (outside of the law enforcement system).

Be a good defender of animals yourself, using welfare and humane treatment. But don't be ashamed of eating them. Who is the dominant animal? Man. That is the way nature is.

Q12. ADDITIONAL INFORMATION

The animal rights people use militaristic terms like "this is a war." If they begin acting out these declarations of war, then they should remember the old maxim that the one who starts the war might be the loser in the end.

There is a trend toward humane treatment. Those are human rights. You are putting into human laws various behavioral codes. These are not animal rights. If the people say we cannot kill endangered species, we get a law for that. This is not an ethical question but a legal question. Breaking this law brings you judicial retribution, not ethical retribution; these are legal questions. The Constitution has nothing about animal rights. Statutes come up, which are new matters of police power. You can assign police powers to animal issues, but then we are talking not about rights, but about police power. "Shoulds" are ethical questions. There is no right answer to such questions; they must be answered by your ethical systems. But these are not rights, this is not a question of rights, because nothing protects that species from other species acting against them. The notion of animals having rights is nonsense. If it is not contained in our Bill of Rights, it is not a right.

David Barbarash

Q1. BIOGRAPHICAL PROFILE

I am an animal liberation activist. I am also an earth liberation activist, and I work against genetically modified organic foods ("gmo"). I am in my mid-thirties and I have (finally) moved out of the city to find some peace and quiet, and live a less stressful life in a rural community. The city offers too many distractions that impede my concentration on work. I am much more concerned now about my own health, as I have experienced the detrimental effects that city life creates, such as increased pollution and stress, and the overall deterioration in the quality of life.

I was born in Montreal in 1964, but moved to Toronto with my family when I was three years old. My father, who worked for the Canadian national radio network CBC at the time, sought a transfer from Montreal, which was experiencing political tensions and violence from the Quebec separatist group FLQ. I lived in Toronto until 1989, when I left to live in the more relaxed climate of Vancouver, BC, on Canada's west coast (I affectionately refer to this as "the year I escaped from Toronto"). I then began several years of travel and activism around North America.

I enjoy the work that I do, which is currently to be the spokesperson for the Animal Liberation Front (ALF). I enjoy the challenge of engaging the media to increase the importance of animal liberation issues in mainstream society. I enjoy speaking out in support of activists who risk everything for nonhuman animals. I enjoy representing the outlaws of our society.

Prior to this role, I enjoyed being one of those outlaws myself. It brought incredible rewards, such as the satisfaction of knowing that I directly saved many animal lives.

I also enjoy other pursuits, such as creating public video events. I started a project called Black Cat Video. I hold monthly screenings of activist and political videos that cover topics that range from animal and earth liberation issues, to globalization and social justice concerns. I enjoy disseminating information to people, be it through video or magazines, because I know that knowledge is power. With knowledge, we, as a civil society, have the power to create massive changes to the injustices we confront in our society.

Of course, I also enjoy nonactivist-related activities. I listen to music (hip-hop, electronic, reggae), watch non-Hollywood films, enjoy the television show *The Simpsons*, hike and camp in wilderness areas, and take long road trips.

Q2. BECOMING INVOLVED

I became involved in animal rights issues through a gradual, but fairly quick process. The first issues I became concerned with were nuclear weapons production and testing in Canada in the early eighties, particular those surrounding the cruise missile. I was already gravitating toward the punk rock movement, so my involvement within the peace movement lead me to the more radical edges, which involved anarchist, antiauthoritarian, and antifascist philosophies. I was involved in several acts of civil disobedience to create awareness around nuclear weapons. The connections I made through that activism led me to people involved with animal rights issues. Fighting the injustices of nuclear weapons and war naturally led me to question, and later fight, the injustices against other species. Besides, I wasn't protesting war only because it affected us; it affects all species and our very planet itself.

Q3. IMPORTANT ISSUES

There aren't any particular issues within animal rights that I spend most of my time on. All the issues surrounding animal abuse, torture, and killing concern me deeply. Nevertheless, in my role as a spokesperson for the ALF, I am limited as to which issues I can address at any given time. For example, recently the ALF liberated quail and ducks from a product-testing lab in Colorado, so I addressed the issues surrounding vivisection and toxicity testing. The next week the ALF released minks from a fur farm in Iowa, so then I addressed issues of animal confinement and torture for vanity clothing items.

I don't necessarily create a hierarchy of issues; all animal abuse concerns me. However, some issues stand out in my mind, because the abuses involved are extremely horrific. Still, when I try to list which issues are the most important, I cannot limit it to one or two or three. There are horrifying things done to animals across the whole spectrum of concerns. Perhaps the question should be, "What are the horrific things that happen to animals in regard to each particular animal abuse issue?" I could go on for pages!

People should care about animal rights issues because what is done to animals is also being done to humans. We feel the repercussions of animal abuse, whether it's cancer and disease from eating animal flesh, or being a victim of a murderer who first practiced on animals as a child. How we treat the animals in society greatly reflects on how we treat each other. Given the violence, disrespect, and callous attitudes we have toward our fellow humans, I believe we need to address the root causes. Those root causes are found in animal abuse issues.

Q4. GOALS & STRATEGIES

I believe that all forms of activism are necessary to achieve the goal of animal liberation. I believe that everything from writing letters to newspapers and politicians and companies, to holding protest signs in front of stores, to marches and nonviolent civil disobedience, to acts of economic sabotage and animal liberation, are all necessary. No single method will work on its own; all kinds are needed to apply pressure and achieve change.

My part in this process is to speak out in support of those people who commit illegal acts, such as destroying the property of animal abusers and rescuing animals. It is hard for people to understand or accept that these tactics are legitimate, even within the animal rights community, so I see my role as extremely vital. To this end, I speak publicly to the media about who these anonymous activists might be (in general terms) and why someone would be so moved to commit these actions.

Q5. DEFINING SUCCESS

The ultimate goal of the ALF, and of myself as a former ALF activist and now their spokesperson, is the complete abolition of all animal abuse in our society. The ultimate goal is the creation of a society that recognizes the inherent rights of all animals, both human and nonhuman, to live their/our lives free from pain, torture, and abuse. I do not believe I will see the fulfillment of this goal in my lifetime, but I believe that I am contributing greatly toward its ultimate accomplishment.

I can say "we have succeeded" when:

- animals are no longer being experimented upon in labs
- the majority of people in the industrialized nations are no longer eating a primarily flesh-based diet, and factory farms are seen as the animal concentration camps that they are
- wearing fur and leather is recognized as being akin to wearing cat and dog fur (or human skin)
- zoos and aquariums cease to exist because the animals held within them have been recognized to have committed no crimes which would justify such cruel imprisonment
- rodeos and animal circuses are abolished for the obvious cruelty they inflict
- hunting and trapping are banned
- dissection in the classroom is recognized as an integral part of desensitizing our youth and is replaced with books and computer programs
- industrial logging, clear-cutting, and road building in our wilderness areas are recognized as destroying not only the forests but also the homes of multitudes of animal species

When all that, and more, is accomplished, then we will have succeeded, and we will have become a society that respects animals as we should respect each other.

Q6. GROUPS & ORGANIZATIONS

I work directly with the North American Animal Liberation Front Press Office, and I work closely with related animal liberation groups involved with producing information or supporting nonviolent protest activity, such as the North American ALF Supporters Group, Frontline Information Service, and the animal rights magazine *No Compromise*. I also work closely with other underground activist press offices, such as the North American Earth Liberation Front Press Office (which speaks for the Earth Liberation Front activists), and the Genetix Alert Press Office (which speaks for underground anti-gmo activists who destroy genetically engineered crops).

Q7. "THE OPPOSITION"

I consider "the opposition" to be those people who have organized themselves to specifically support the continuation of animal abuse. I don't consider "normal" everyday people our opposition; rather I consider regular folks to be capable of understanding, learning, and compassion. The meat eater, fur wearer, consumer of animal products: these are all people who I believe are capable of change. Once we understand the issues of cruelty we can begin the process of eliminating it from our lives. I don't believe people in general are evil. However, those people who have vested interests in continuing their involvement in animal abuse, who stand to profit greatly from the continuation of animal abuse, and who actively oppose people who wish only compassion and peace for nonhuman animals are the opposition, and they will be fought every step of the way.

Q8. HEROES

The people who inspire me the most are the regular people on the front lines battling to save animals from torture, battling to save our old growth forests from destruction, and battling all forms of oppression against animals, humans, and our planet. The people who risk everything—their liberties and freedoms, and the comfort they could easily choose to have in our society by ignoring these issues—by putting themselves out on the line to selflessly help others who cannot speak out for themselves are my heroes.

I don't wish to name any particular people as "heroes." Part of what we, as a conscious community of activists, are trying to move away from is this notion that some people are more valuable or greater than others (and, by extension, that some species (humans) are more valuable than other species). The people I consider heroes are regular, normal, everyday people, and by naming individuals we put them up on a pedestal to be admired and gawked at. By doing this we encourage the idea that it is only extraordinary people who are capable of animal liberation activities. This is definitely not the case. Anyone can be a "hero" by simply taking a stand and refusing to participate any longer in animal cruelty.

Q9. RELIGION/SPIRITUALITY

I consider myself to be a very spiritual person, and my spiritual beliefs are an integral part of my life and the activist work that I do. I rejected the standard Judeo-Christian belief systems at an early age, even while I was enrolled in a Jewish school, and I gravitated toward a more earth-centered spirituality. Wicca and Paganism were interesting to me because of the connections they made between us as humans and the earth and all her critters. As an animal and earth liberation activist, integrating earth-based religion into my life seemed only a natural progression.

This integration has helped me greatly in my activist work. It has led me to a deeper understanding of the interconnectedness between all life on Earth, how we are all part of the intricate web of life. This understanding is a big part of what fuels my rage against animal abuse. When one part of the web is attacked, it compromises the entire structure. By saving animals and protecting the earth, we will find our own salvation, literally and metaphorically.

I've also benefited from my beliefs in times of great stress, such as when I was imprisoned for my activism. My strong faith in the belief that I was a part of something larger, something beautiful and amazing, what is commonly referred to as "life on planet Earth," has kept my mind strong under stressful circumstances.

Q10. FUNNY/STRANGE EXPERIENCES

I'm always taken aback when I find support for the work I do come from the most unlikely of places. When I was being transferred from California to Edmonton, Alberta, to face charges related to liberating twenty-nine cats from the University of Alberta, I was held overnight at an RCMP detachment in Richmond, BC. The officer on duty that night was an older guy, very close to retirement. He came up to me and asked, "Are you that guy who took those cats?" Being cautious of some kind of entrapment my answer was noncommittal, but my smile said "Yes!" He smiled back, gave me the thumbs-up, said "good job!" and walked away.

Q11. LESSONS/ADVICE

There would be both practical and philosophical advice I would give to the new animal rights activist. Philosophically, I would advise the person to study the different levels and options of activism that people employ. Understand that there is no "one way" to reach the goal of animal liberation—that it will only be attainable through the multiple and varied tactics people take part in. The armchair activist letter-writer is fulfilling as important a role as the volunteer at the animal shelter, as the protester in front of the fur store, as the unknown activists who break the cages and frees the animals. I would advise them to direct their energy, anger, and enthusiasm toward the animal abusers, and to not get caught up in condemning those activists with whom they disagree (on a tactical level). I would advise them that the first most important thing they could do would be to eliminate animal products from their life and become vegan.

Practically, I would advise the new animal rights activist to think long and hard about whether they might be more interested in radical nonviolent direct action. If an activist is seriously considering this honorable path, then he or she needs to consider many other practical life changes and issues, such as the need for complete security, the decision to not get arrested at any above-ground protest, and to not even attend any public protest events.

Q12. ADDITIONAL INFORMATION

I'd like to end by talking about the word "terrorism." This is a label that is loosely bandied about by those who are involved in animal abuse or in law enforcement to describe animal rights activists. Typically it's used to whip up fears against the ALF because they break the law to free animals, and they destroy property. But just who is it committing "terrorism," and who is being terrorized?

In the labs we have animals in restraints being cut open while still alive; electrodes, chemicals, drugs, and poisons are routinely applied. Animals in fur farms are caged in tiny cages for their short lives and then killed by anal electrocution or neck breaking. Animals in circuses are beaten and whipped to perform unnatural acts. The list is endless. This is the real terrorism, and it occurs every day nonstop.

The ALF are people who give freedom to these animals and who destroy the property that is used to torture and abuse these animals. The ALF abide by a strict guideline of nonviolence—no animal or human has ever been hurt or killed in the course of an ALF action. These people are not terrorists; these people are genuine freedom fighters. URL: *http://www.animalliberation.net/media/naalfpo.html.*

Don Barnes

Q1. BIOGRAPHICAL PROFILE

I was born in Colorado Springs, Colorado, on September 1, 1936, in a little tiny house on a very steep hill. It was a nasty winter, and the doctor never made it back to that little tiny house on the very steep hill to circumcise me, thereby saving me from this entirely unnecessary mutilation. As a consequence, I have enjoyed a very happy sex life for the past fifty years.

I also enjoy billiards, public speaking, vegan cooking, writing, and saving nonhuman animals.

Q2. BECOMING INVOLVED

See my chapter in the book *In Defense of Animals* by Peter Singer.

Q3. IMPORTANT ISSUES

I spend a lot of time supporting the Texas Snow Monkey Sanctuary near Dilley, Texas. I sometimes arrange for homeless monkeys (snow monkeys, vervets, baboons) to come there to be released in large (five- to sixty-five acre) outdoor enclosures, so that they can interact meaningfully with their fellow species. I arrange for free fruit and vegetables from San Antonio produce outlets; I also rebuild trailers and pens, rescue rattlesnakes, help clean and feed, and so on. In the process, I interact with zoos, other sanctuaries, and those involved in sanctuary accreditation.

I spend time releasing trapped possums, skunks, raccoons, and other animals in my local area. I help rescue birds, snakes, mice, and rats.

I work on "problems" with the overpopulation of white tail deer.

I speak and debate the merits (or lack of) of animal experimentation at universities and colleges.

I am very active in trying to protect dolphins from being trapped and placed in concrete prisons for human amusement. To that end, I work closely with folk from The Dolphin Connection and Dolphin Embassy U.S.A. in Corpus Christi, Texas. This work puts me in contact with dolphin advocates around the world.

Q4. GOALS & STRATEGIES

Short-term goals are difficult to define. Examples of activities I might participate in are watching a trapped skunk scamper his or her way to freedom; setting up and maintaining a colony of feral cats at a university; helping one person at a time eschew the consumption of animal products; giving an interview for a television crew on the evils of circuses or rodeos or the fur business; rescuing an abused animal; finding a companion horse for one housed alone; disrupting a hunt, thereby saving the hunted creature; feeding tomatoes (peanuts, watermelon, carrots, etc.) to 350 hungry monkeys; or helping to affect an attitude change in a prejudiced ("speciesistic") human.

Long-term goals are easier to define, though more difficult to realize. I try to educate, to bring about a nonspeciesistic world.

Obviously, I do all of these things. I am convinced that the plight of nonhuman animals will not be ameliorated until there's a major alteration of the *zeitgeist*, not only in the United States, but all over the world. I speak to civic groups, churches, and public gatherings of all kinds. I do radio and television interviews and organize and participate in protests. I work with local animal control people, city council members, and the mayor of San Antonio on the overpopulation of dogs and cats.

I refer local citizens to other agencies when they call me with questions about nonhuman animals. I administer a rescue fund through a local organization and work closely with local veterinarians to help injured or abused animals.

I also participate in cooking contests (chili), teach vegan cooking, and try to educate restaurateurs about the joys of veganism.

Q5. DEFINING SUCCESS

We will never succeed. Humans will overpopulate the world, and the earth will eventually succumb to that pressure. We can only hope to delay that inevitable conclusion.

Q6. GROUPS & ORGANIZATIONS

I am the southern field representative for the Animal Protection Institute. As such, I monitor Texas, Louisiana, Oklahoma, Florida, New Mexico, and Arizona for opportunities to intervene in a meaningful way for nonhuman animals.

I am the executive director and treasurer of VOICE for Animals, a local volunteer animal rights organization of eleven years.

I also work as the southern regional representative of Vegan Outreach, and help distribute the pamphlet "Why Vegan?" to hundreds of outlets.

Q7. "THE OPPOSITION"

People who maintain a wide breach between their species and all other life forms.

Q8. HEROES

Tom Robbins, Don Quixote.

Q9. RELIGION/SPIRITUALITY

I am a devout atheist. Due to the human fear of mortality, humans conjure a life after death for those created in "god's" image. Most religious texts reinforce the schism between humans and nonhumans and justify our domination of others. "Thou shalt not kill" applies only to other humans. The cruelty toward other animals I have experienced in my lifetime prevents me from believing in a compassionate godhead.

Q10. FUNNY/STRANGE EXPERIENCES

I did not view this directly, but it deserves mention. At the Texas Snow Monkey Sanctuary, some eighty miles south of San Antonio, some 350 snow monkeys roam freely in a sixty-five-acre enclosure surrounded by an electric fence. They swim, teach their offspring to swim, to respect the rules of the troop, and so on. There are three Javelina (wild pigs) living within this enclosure, obviously inside the area when the fence was built and activated. They get along well with the monkeys, as well as share their food and water.

A few weeks ago, Tom Quinn, the site manager, spotted a macaque grooming someone. As he got closer, he saw that this monkey was grooming one of the Javelina. As he watched, the monkey leapt on to the back of the pig, bouncing up and down like Gene Autrey on his horse. The Javelina remained passively standing until he spotted Tom at which time the episode ended.

I wonder: Just what kinds of interactions between nonhuman beings really happen in the wild? If a monkey grooms a pig and no one sees it, did it happen?

I could probably write an entire book about unusual (funny or not) things that I've observed in my twenty years in this movement. For example, I was in attendance at a protest at the National Institute of Mental Health some ten or twelve years ago. A lot of the activists had sat down and blocked the entrance. The cops arranged for a bus to come take the arrested activists away. One of the "leaders" of the protest—the executive director of a large West Coast animal rights organization—was busy doing an interview with the press, when he noticed that the bus was about to leave for the police station. He ran to the bus, yelling, "Wait! Wait! I want to be arrested too!"

What some people won't do for that "red badge of courage"!

Q11. LESSONS/ADVICE

To those opposing animal rights activism, I would say, "Get a life! Open up yourself to an understanding that all life-forms are interconnected and life on earth is dependent upon maintaining the symbiosis of natural life."

To the beginning animal rights activist, I would say, "Go slowly until you understand where the critical pressure points are, and then throw all your weight toward change."

I would also advise them not to idolize others in the movement, not to follow without understanding where the path might lead. I would ask them to look to themselves for their own motives and to refrain from using the movement for their own personal ends. I would caution them against prioritizing species, projects, organizations, or individuals and to accept that the "little old lady in tennis shoes" is as valuable to this movement as the most renowned philosopher. This movement suffers from egos and self-interests; I would ask them to keep the goal in mind, for example, the minimization of stress, pain, suffering, and death for all animals, human and nonhuman.

Q12. ADDITIONAL INFORMATION

Don't take yourself too seriously; you're simply one animal on a very large planet.

Gene Bauston

Q1. BIOGRAPHICAL PROFILE

I am a human being who is disappointed with our species' lack of concern and compassion about other species on Earth.

I was born in Los Angeles, California, in July 1962. I was encouraged to get a well-paying job and to strive for the American Dream, that is, rich in material goods. I never felt comfortable seeking comfort when so many others starved and suffered. I have always been inspired by animals and nature. Having grown up in southern California, I particularly like the beach at sunset, where the Earth meets the ocean, and the day meets the night. This seems to be a transitional place and time, and I believe that each day presents an opportunity for each of us to learn, grow, and be transformed. The only constant is change.

I also enjoy playing and watching sports: the human drama of athletic competition.

Q2. BECOMING INVOLVED

It was a gradual process punctuated and strengthened by individual events.

My first glimpse of the negative impact that humans have on animals involved a deer that was injured in the backyard of a neighbor in the Hollywood hills, where I grew up.

My first insight into the ugliness of meat consumption occurred during my high school years, when I came home and saw a chicken body, not a meal, on the stove.

Q3. IMPORTANT ISSUES

I spend much of my time trying to prevent the suffering of animals that are exploited for food production, particularly the suffering of "downed animals" and those kept in intensive factory farming confinement. I also encourage humans to adopt a vegan lifestyle. Both of these objectives are good for all animals, including humans, whose hearts and arteries are hardened by factory farming and the consumption and use of animal products.

Q4. GOALS & STRATEGIES

In the short term, I hope that cruel factory farming practices can be outlawed, increasing public awareness and leading toward a vegan world.

I research, write, organize, lobby, and work to educate the public on these issues. When it is not possible to reason with those who cause injustice and cruelty, I will protest.

Q5. DEFINING SUCCESS

I believe that we can achieve various short-term successes (e.g., prohibitions on specific forms of factory farming confinement, mutilations, and inherently cruel production practices), and I hope that we can inch our way toward a world in which no human intentionally behaves in a way that causes other beings to suffer. I am not sure that we will ever "succeed" in preventing all suffering, but I believe that we should strive toward this ideal.

Q6. GROUPS & ORGANIZATIONS

I am cofounder of Farm Sanctuary, and I work closely with this organization. Farm Sanctuary is a national nonprofit organization that works to stop farm animal suffering through direct animal rescue and care, investigative campaigns and exposés, legislation and litigation, and through public education and awareness efforts. Our website is *www.farmsanctuary.org*.

I also work with many other groups to achieve our shared goals.

Q7. "THE OPPOSITION"

I believe that everyone who has an open mind and a desire to improve the world is our ally.

Our opposition is those people who believe that change is unattainable and that compassion is unimportant. Our most difficult opposition is the group of organizations who are economically and emotionally invested in the exploitation of animals for profit. We urge these people to make significant changes, and change can be difficult. But I believe that such changes will ultimately benefit everyone.

Q8. HEROES

Gandhi, Thich Nhat Hahn, Martin Luther King Jr., and Nelson Mandela.

Q9. RELIGION/SPIRITUALITY

I am not a particularly religious person, though I grew up in a very Catholic family. In my opinion, historically, religion has tended to hinder compassion to other animals because it separates humans from other animals. Religious leaders also tend to be judgmental and intolerant of other viewpoints, and I am deeply opposed to such intolerance. Ultimately I believe that religious institutions can evolve and broaden their capacity to encourage compassion beyond "the chosen people" and beyond the human species.

I believe that connecting with animals and nature can be a religious/spiritual experience.

Q10. FUNNY/STRANGE EXPERIENCES

[No response.]

Q11. LESSONS/ADVICE

As animal advocates, we are faced with an impossible situation. Because we cannot immediately stop all animal cruelty perpetrated by our society, we are forced to tolerate the intolerable. It is natural and appropriate to feel hurt and angry about animal suffering, but we should strive to channel our pain and anger into a positive response, guided by respect and compassion.

Listen carefully to others and try to understand their points of view, even if we disagree vehemently with their actions and statements. We should strive to "love the sinner but hate the sin." We cannot control others' behavior, only our own, so we should strive to behave as humanely and respectfully as we can.

Q12. ADDITIONAL INFORMATION

[No response.]

Marc Bekoff

Q1. BIOGRAPHICAL PROFILE

I am a human being first and foremost. By profession, I am a professor of Biology at the University of Colorado at Boulder. I was born in Brooklyn, New York, on the 6th of September 1945. I grew up in a very warm, compassionate, and loving household, and always felt very close to non-human animal beings ("animals").

I enjoy numerous activities. I love my "work," and I do not really think of it as "work." I ride my bike 8,000–12,000 miles a year, love to read spy novels and mysteries, and enjoy all sorts of music. I love spending time with my companion dog, Jethro. He is a true hero in that he is incredibly compassionate and constantly reminds me of the power of compassion and reverence for all life.

Marc Bekoff.

I am a biologist, a lover of the diverse and wondrous life on this splendid planet. As a scientist who has been lucky enough to have studied social behavior in coyotes in the Grand Teton National Park in Jackson, Wyoming, the development of behavior in Adélie penguins in Antarctica near the South Pole, and the social behavior of various birds living near my home in the Rocky Mountains of Colorado, I have learned a lot about these amazing animals (and many others). I am very concerned about what humans are doing to other animals, and to the planet in general. While some of my views may make it seem as if I want to stop all animal research, including my own, and the use of all animals everywhere, this is not so. I am just not very happy about what is happening to the wonderful animals with whom I am privileged to live and share the earth. Are you?

Q2. BECOMING INVOLVED

My interests in the lives of animals, and how they are treated by humans, seem to be innate. I never really deeply thought about these issues, yet I always had deep feelings about them. My parents, who still cannot figure out how I came to this profession, tell me that I have always "minded" animals. Although I was not raised with animals, I used to ask about what they might be thinking or feeling as they went about their daily activities. In a nutshell, the phrase "minding animals" means caring for them, respecting them, feeling for them, and attributing minds (mental states and content) to individuals. Since I began working with animals, I have always spent a lot of time pondering non-human/human relationships. I am often upset by the terrible things that humans do to other animals with whom we share Earth. I recognize fully that many people who harm non-humans for purposes of research, education, or amusement also bring some joy to some animals at other times. Of course, non-humans do not always suffer at the hands of humans. I find myself focusing on horror stories not because I am a pessimist (whose glass is always half-empty), but rather because it is more important to call attention to the incredible pain and suffering of animals at the hands of humans, than it is to remind people of the good things that are done by humans for non-humans' benefits.

My early scientific training as an undergraduate and graduate student was grounded in what the philosopher, Bernard Rollin, has called the "common sense of science," in which science is viewed as a fact-gathering, value-free activity. Of course, science is not value-free. We all come to our lives with a point of view, but it took some time for me to realize this truth, because of our heavy indoctrination (and arrogance) concerning the need for scientific objectivity. In supposedly "objective" science, animals are not subjects, but objects that should not be named. Close bonding with them is frowned upon. However, for me, naming and bonding with the animals whom I study is one way to show respect for them.

With respect to the plight of the non-humans who were used in classes or for research, there was little or no overt expression of concern for their well-being. Questions concerning morals and ethics rarely arose. When they did, these questions were invariably dismissed either by invoking self-serving utilitarianism (I call this "vulgar" or "facile" utilitarianism), in which suspected costs and benefits were offered only from the human's point of view (with no concern for the non-human's perspective); or by simply asserting that the animals really didn't know, care, or mind (or whatever word could be used to communicate the animal's supposed indifference to) what was going on. Only once do I remember someone vaguely implying that something beneficial for the animal might come out of a research project.

Here is a story of some events that changed my life. One afternoon, during a graduate course in physiology, one of my professors calmly strutted into class announcing, while sporting a wide grin, that he was going to kill a rabbit for us to use in a later experiment by using a method named after the rabbit himself, namely a "rabbit punch." He killed the rabbit, breaking his neck, by chopping him with the side of his hand. I was astonished and sickened by the entire spectacle. I refused to partake in the laboratory exercise, and also decided that what I was doing at the time was simply wrong for me. I began to think seriously about alternatives. I enjoyed science (and continue to enjoy doing scientific research), but I imagined that there must be other ways of doing science, that would incorporate respect for animals, and allow for individual differences among scientists concerning how science is conducted.

I went on to another graduate program, but dropped out because I did not want to kill dogs in physiology laboratories and cats in a research project. My research centered on vision in cats. I truly enjoyed the challenge of determining how cats see their world. However, once the killing of experimental animals began, I also truly hated killing the animals to localize lesions in various parts of their brains. Recently I learned that the famous biologist, Charles Darwin, might also have left medical school after one year, because he was "repulsed" by experiments on dogs. In his book, *The Descent of Man*, Darwin wrote the following about those people who experimented on dogs: "this man, unless he had a heart of stone, must have felt remorse to the last hour of his life."

One morning I woke up very disturbed about the whole thing, and decided that I could not continue to kill the cats. The eyes of the cats especially tormented me as they were being prepared to be killed. What an undignified end to a life! I simply didn't want to kill animals as part of my research. I could not justify this murder using any form of utilitarianism. I (and three medical students) refused to partake in some physiology experiments that used dogs, and to our amazement, we were excused without prejudice from doing so, although the distinguished professor couldn't understand why we didn't want to kill the dogs; he asserted that they would have died in an animal shelter anyway. To his credit, though, he remained true to his word and I applaud his permissiveness and his open-mindedness. At the end of the term, despite these reprieves, I left this program because I could not do the research that I wanted to without killing animals or being responsible for their deaths. If I must forego learning something because I must kill animals to do so, then so be it.

Q3. IMPORTANT ISSUES

The issues on which I spend a lot of time include the horrible lives of animal prisoners in zoos, wildlife theme parks, aquariums, and research laboratories. I am also very concerned with the use of animals in education: I am a strong opponent of dissection and vivisection in the classroom. These are the issues closest to my heart.

I want people to learn about the awesome and magnificent lives of "wild" animals, not their captive relatives, whose impoverished lives in zoos are seriously compromised. They have no freedom of choice; they cannot go where they want to go. People should care about these issues, because zoos promote and represent false images of animals.

My interest in the lives of research animals stems from my own research; I know how horrible their lives are. The arrogance of some of my colleagues also motivates me to question their goals and the way they treat the animals (on whom their lives and reputations depend).

Furthermore, many biomedical models that stem from animal research simply do not work toward those goals for which they are intended. Unfortunately, the use of animal models often creates false hopes for humans in need. It is estimated that only 1–3.5 percent of the decline in the rate of human mortality since 1900 has stemmed from animal research. The prestigious publication, the *New England Journal of Medicine*, called the war on cancer a qualified failure. More than 100,000 people die annually from the side effects of animal-tested drugs. Early animal models of polio also impeded progress on finding a cure. As pointed out by the Medical Research Modernization Committee, Dr. Simon Flexner's monkey model of polio misled other researchers concerning the mechanism of infection. He concluded that polio only infected the nervous systems of monkeys; but research using human tissue culture showed that poliovirus could be cultivated on tissue that was not from the nervous system. Chimpanzees were used to study AIDS, but chimps do not contract AIDS! Many people die from biomedical models because the diseases produced in animal research are artificially induced, while the naturally occurring course of the disease is quite different.

I also continue to develop my notion of deep ethology. I use the term "deep ethology" to stress that people are an integral part of nature, and they have unique responsibilities to nature. Deep ethology means respecting all animals, appreciating all animals, showing compassion for all animals, and feeling for all animals from one's heart. Deep ethology also means resisting speciesism. Our respect for animals does not mean that we can then do whatever we want with them!

Our starting point should be this—we will not intrude on other animals' lives unless we can justify an override of this maxim: that our actions are in the best interests of the animals, irrespective of our own desires. When unsure about how we influence the lives of other animals, we should err on the side of the animals.

Some guiding principles include:

1. putting respect, compassion, and admiration for other animals first and foremost;
2. taking seriously the animals' points of view;
3. erring on the animals' side when uncertain about their feeling of pain or suffering;
4. recognizing that most of the methods that are currently used to study animals, even in the field, are intrusions on their lives, and thus, exploitative;
5. recognizing how misguided are speciesistic views, using very vague notions of intelligence and cognitive/mental complexity for determining assessments of well-being;
6. focusing on the importance of individuals;
7. appreciating individual variation and the diversity of the lives of different individuals in the worlds in which they live;
8. using common sense and empathy, which some say have no place in science; and
9. using broadly based rules of fidelity and non-intervention as guiding principles.

Q4. GOALS & STRATEGIES

I regularly write, educate, conduct research, and protest the mistreatment of animals. These are short- and long-term strategies. Some problems simply need a lot of time for their resolution.

My goal is to have people become more compassionate and respectful to all animals and the inanimate environment. We should view animals as our friends and partners in our effort to make the world the best it can be, for them and for us. We need to love animals and Earth.

Being concerned about animals does not mean we are "insensitive to humans." The two are not remotely related in my view. For example, stopping animal research would mean that we would develop better models and cures for human diseases.

Q5. DEFINING SUCCESS

I feel success will come when animals are no longer eaten, imprisoned in zoos, used in research, cut up in educational pursuits, and otherwise exploited by humans for their own anthropocentric interests.

Q6. GROUPS & ORGANIZATIONS

While I tend to work on my own, I also have written and worked with a number of animal protection organizations including the Animal Protection Institute, Rocky Mountain Animal Defense, Medical Research Modernization Committee, Jane Goodall Institute (especially its Roots & Shoots Program), and *Animals' Agenda*.

Q7. "THE OPPOSITION"

Any group that routinely puts humans "above" other animals, with no regard for their lives, are the opposition.

Q8. HEROES

Jane Goodall and Henry Spira are among the people who inspire me the most. They are not heroes, but sources of deep inspiration.

Q9. RELIGION/SPIRITUALITY

Religion does not play an important part in my life, but different forms of spirituality do. I feel spiritually connected to animals, and feel that my deep empathy and compassion for their lives enables me to sense more about their lives, to feel qualities—who these amazing beings are—that others might not be able to feel deeply in their hearts.

Q10. FUNNY/STRANGE EXPERIENCES

One day I was riding my bike with my friend, Brad Wallace, when we saw a very young bunny running about, trying to dodge traffic and jump up onto a curb. He couldn't make it because he was so small and the curb so high. We stopped in the middle of the road, laid our bikes down in front of oncoming rush-hour traffic, and ran after the bunny to save his life. At first people were angry, but when I told them what we were doing, some joined in. The tolerant commuters began yelling at the irate drivers to cool their jets.

I eventually caught the bunny, put him in the back pocket of my cycling jersey, and carried him to a nearby field where I gently placed him down. He looked up at me as if to thank me and ran off. Some of the commuters applauded as we rode off. I want to believe that this event actually changed some of the people's minds and hearts about animals.

Q11. LESSONS/ADVICE

One must have a passionate vision. I would say that the most important starting point is to believe that it is always wrong to harm other animals intentionally: that humans are obliged to honor the lives of nonhuman animal beings even if it would benefit us not to do so.

I also advocate patience. Telling others what to do never works. Long-lasting changes are more productive than short-term changes. The changes must be changes of the heart—deep changes—and not superficial changes that are temporary or ephemeral.

Q12. ADDITIONAL INFORMATION

After editing *The Encyclopedia of Animal Rights and Animal Welfare* (Greenwood Publishing Group, 1998), I realized how much work is left to do in the area of animal protection. I hope to expose more youngsters to the critical issues in my children's book *Strolling with Our Kin: Speaking for and Respecting Voiceless Animals*. I develop many of my ideas in two recent books, *Minding Animals: Awareness, Emotions, and Heart* and *The Ten Trusts: What We Must Do to Care for the Animals We Love* (with Jane Goodall). In June 2000, Jane and I founded the organization Ethologists for the Ethical Treatment of Animals/Citizens for Responsible Animal Behavior Studies (*www.ethologicalethics.org*).

Brian Bishop

Q1. BIOGRAPHICAL PROFILE

I am described by some as a political Magellan who went around the world to the left and is coming in from the right. If you threw in libertarian tendencies, maybe you could say I took the trip via a polar route. I was born in 1957 into the "ivy league" section of Providence, Rhode Island, which is culturally and intellectually dominated by Brown University. As a consequence, I was raised in an environment that was steeped in questioning authority. Ironically, this mantra of individualism was "group think" during the youth movement era. One could thus question authority without deciding whether the preference was for a less controlled laissez-faire lifestyle or a more "benign" authoritarianism (if there is such a thing).

While I credit the liberal environment in which I matured for fostering those concepts, it did so during a relatively conservative time. Thus, I see in hindsight, it was not the values of liberalism, but the out-group status to which I was attracted. Now that our country has a relatively liberal outlook, I find that there is far less scholarly interest in the Brown community to critically examine the "great society," or the vast environmental programs, that came when political power passed to my mentors and co-conspirators in the "youth movement." Admittedly, this is a critique of higher education by someone who rejected that path. I graduated from high school in 1975 and haven't seen the inside of a classroom again, except for a few short stints teaching Math in various private schools.

Q2. BECOMING INVOLVED

Like many caught up in the antiwar movement, I saw the next great challenge as the environment and went "back to the land": moving south of Providence to an area known for drunken woodcutters and inbreeding. I fit in reasonably well. When I got there, I found that the folks who saw the environment as a political cause had, ironically, become a roadblock to those of us who saw it as a personal challenge. I faced interminable conflict with my former brothers-in-arms, in the person of the local planning board and the state environmental agency.

I actually took the "work within the system" approach for some years, believing that all the head-butting I experienced with government was the result of a communications problem, or perhaps a hearing deficit. I said I wanted to "paint my own canvas," and they thought I said I wanted to "paint Kansas." But the system is carefully crafted to block or absorb any divergent viewpoints, and when I stood on the edge of financial failure for trying to defend my style of environmentalism against the bureaucratic version of environmentalism, I was driven to political activism.

While environmental and animal rights advocates might be properly viewed as members of the same set—they are driven to political action by frustration with their inability to effect change on a personal scale—that is where similarities between an empirical environmentalist and organized environmental causes cease. My activism was aimed where it always had been: at questioning authority, getting government out of my life. Ecological activists had the diametrically opposed goal of getting government into my life and establishing a mandatory system of values in relating to animals and the environment.

My firsthand experience of raising and slaughtering goats, pigs, and chickens, as well as my interaction with the hunting community, gave me a growing respect for sustainable use for both wild and domestic animal populations. So I did engage in serious thought to arrive at a personal value system for animal use, but it is decidedly at odds with the system that animal rights proponents seek to mandate. As a dedicated omnivore, I considered personal involvement in the butchering end of farming as the Rubicon to be crossed in my own values system. I did not conceive of this as some necessary test to justify meat eating by others, but as a part of enhancing my own overall environmental ethic.

I do not consider it a fault, but an unavoidable reality, that the vast majority of our culture is disconnected (for obvious reasons) from the semi-subsistence experiences to which I exposed myself. I did not see them as a necessary trial by fire in order to justify civilization. In my view, such introspection is not a collective responsibility. Each person justifies his/her own existence and is free to either take preprogrammed responses (to issues that they don't wish to experientially confront), or to strike out on their own.

Early in my activist career, I was to learn a lesson about adopting the views of perceptually "pro-environment" groups when I gained firsthand experience at a flashpoint over environmental policy, and thus raised my horizons from battling merely at the local level, to battling environmental and animal rights interests across the board. I was taken in, hook, line, and sinker, by a national environmental propaganda campaign.

Even while I was beginning to doubt the veracity and good faith of environmental activists in my local setting, like all well trained progressives, I heeled patiently in line to buy tuna that had a happy little dolphin on the can. Swept up in the *Heaven Can Wait*–spawned consumer environmentalism, I was right there with Warren Beatty's character when he scoffed at any economic considerations in limiting bi-catch; "we don't care what it costs, we care what it makes." This is a ditty which has since proved to be a poor business model, but at the time seemed like wisdom which, correctly paraphrased, could form the basis of the 11th commandment.

Thus, this longhaired, bearded, leftover hippie who had become a property rights advocate (as a consequence of government interrupting his trip back to the land) was still carrying plenty of progressive water at this point.

This seeming dichotomy of embracing national environmental campaigns while jousting with their implementation at the local level was not to last. When my outspoken opposition to local environmentalists brought me to the attention of property rights and multiple use advocates in other states, the inevitable interaction led me to question more environmental shibboleths, which before had seemed unrelated to my localized conflict.

It had not been preordained that I become a national proponent of sustainable use. In fact, if anything, I thought that environmentalists like me were only being attacked because they mistook me for some radical western "sagebrush rebel." I quickly found, however, that efforts to keep my property rights were consistently reported to be part of a growing conspiracy that was loosely termed "the backlash," and eventually gained the proper name "Wise Use."

During this period, I still believed myself to be an environmentalist. I would cite my lifelong penchant for recycling (now tempered by an understanding of materials life-cycle analysis which suggests recycling is not always the most conservation-oriented practice), which I picked up as a Boy Scout. I would cite my allegiance to what I presumed to be honest campaigns like saving dolphins from evil tuna fishermen. But, I was alarmed to find that newspapers would not print much of what I said because I used the words "property" and "rights" in the same sentence; while the media seemed to have implicit reverence for the leaders of environmental groups. Thus I was labeled as a Quixotic kook, but a clever one, advancing a corporate-inspired anti-environmental agenda.

I may have been naive in my approach to the press and political activism, but I wasn't slow on the uptake. I decided that if I was a member of some conspiracy I had never heard of, and could not shake this image, I might as well go meet my co-conspirators. Thus, the environmentalists drove me into the waiting arms of the wise-use movement, which had an exceeding paucity of sagebrush rebels from east of the Mississippi.

Serendipitously, one of the first people I met was a woman whose family lost their fishing boat when I (and many other "sheeple") began buying tuna with the little dolphin on the can. No one ever told me that the American fishermen, who pioneered the purse seine tuna fishery by encircling dolphins in the late 1950s, had begun developing ways to release the dolphins from day one. Then I see observer data from the international body overseeing the fishery that the release rate is around 98%, and that there is no "resource" issue in terms of dolphin populations.

Now, I'm really getting mad. If anybody had told me that we released that proportion of dolphins, that Americans pioneered the release techniques, and that American laws and consumers' attitudes were effectively killing our own fleet (while inviting a potentially unregulated international harvest), I would have bought cans that didn't have the dolphin on them! I had been given a touchy-feely animal as a mascot for a supposed environmental cause and had swallowed the bait. I had screwed up the resource economy of our country without accomplishing anything, except for increasing the price of tuna fish.

The next sustainable use proponent I met was an Icelander, known as a thorn in the side of the International Whaling Commission. Here was a guy saying that we should take up whaling again! Was it true what he said: that Minke whale stocks were healthy, and that the

IWC was no longer a resource manager, but an animal-rights proponent? I met a real life version of Crocodile Dundee who had pursued research to support crocodile recovery in Australia, and now worked to sanction crocodile hunting and farming, and international trade in products of the hunt. I met representatives of the Communal Areas Management Programme for Indigenous Resources (Campfire), who advocated elephant hunting as a management tool and income stream to supplement traditional subsistence in Zimbabwe. I met trappers and fur-growers. I saw more mainstream footage of seal hunts. Essentially I was introduced to sustainable wildlife management in the context of many contentious controversies around the globe and by seeing these differently than they had been publicly presented by animal rights moralists, I became an opponent of animal rights philosophies.

Q3. IMPORTANT ISSUES

Any duties that man owes to the animal kingdom arise not from some compact with the animals, but rather as a result of a compact among men. Thus "animal rights" are not, as such, of the animal, but rather are human rights in the sense that human concerns for animal welfare may constrain human freedom of action.

Thus animals have no rights. Our concerns expressed as laws and customs protecting animal welfare define what we owe to one another as humans in our individual treatment of animals. It is not as if there is some abstract principle governing the human/animal relationship that existed before human civilization. The preposterous advancement of lawsuits on behalf of animals as parties in our courts, and the attempts to legislate an animal rights agenda, have permanently hardened my concerns about environmental and animal rights activists.

The implication of ecological activism is that some moral principle of the universe, independent of human philosophy, should guide our resource interactions. I can think of few more self-effacing metaphysical dead ends.

Any form of this thinking, legislated into the social compact, effectively renders the compact invalid. Few greater threats to human liberty exist than environmental and animal rights activism. I can understand how some of these people consider themselves to be out-group pioneers, much as I was, at one time. Nonetheless, the hazard to the social compact and to its guaranteed freedoms is inherent in this increasingly frustrated environmental and animal rights activism.

Many activists have now cast aside long-accepted prohibitions on violence and property destruction. There is a dangerous philosophical underbelly to these ecological movements, which troubles me. Some still labor under the false belief that there are universal moral principles existing independently of human consciousness, and that these new principles can defend their own extreme actions. Chief among these in stature and mainstream appeal is E.O. Wilson. Positioning himself as a Renaissance man for a modern day, he has written an entire book devoted to bringing moral questions under the auspices of scientific authorities. This is as bizarre and wrongheaded an idea as the subjugation of scientists such as Galileo to moral or spiritual authority.

That such a proposition can even be advanced without ridicule is evidence of a danger that will command my lifelong activism in opposition. While we can debate scientific questions that bear on moral principle, few would concede that if scientists concluded that life effectively began at conception, that nations should somehow command an end to abortion. Some people would certainly hold that view, but they would not be entitled to visit that interpretation upon society as a whole.

The notion that ascendant technocrats may readily alter the fabric of human rights is anathema to the human civilization. Even elected officials, who may have great impact on the observance of human rights during their tenure in office, have relatively little influence in weaving the fabric of rights itself. I would afford far less influence than even this to nonelected scientists who purport to govern, at least in a moral sense, as a consequence of their "expertise." This is not to suggest that pursuit of science is a rejection of one's political status, but rather that it should not raise that status above the political standing of nonscientists; and, in fact, should be recognized as a conflict. "Industry" scientists are immediately presumed to have some nonscientific bias toward their interest groups, but scientists representing so-called "public interest" points of view are not viewed as clouded by a similar conflict. The entrance of endless groups of scientists into the discussions without this understanding does serious damage to the integrity of the political process that provides the protection of our rights.

Q4. GOALS & STRATEGIES

I have learned that government steers somewhat like a supertanker. You keep your hand on the wheel for a decade and you can look back and see a slight curve in the wake. There are rare exceptions, but I don't expect that I will experience one. Rather I see that I must write, research, and demonstrate in any way that I can, to guard in perpetuity against the concepts of animal rights which are violative of human rights. I don't hold this out as a selfless endeavor; it is truly a defense of my personal rights that can only be effectively defended by defending everyone's rights. Thus, my strategy is to remain involved and relevant to the debates—keeping a hand on the wheel.

Q5. DEFINING SUCCESS

It is unlikely that I will ever see myself as having succeeded. I can gain small moral victories such as chastening an EPA official here or a politician there. When I first became politically active it was with the simple intent of redressing my own grievances. I had every intention of returning to private life as soon as my problems were fixed. The reality is that changing anything in the political realm is the most challenging task that I have ever undertaken. My original problems are over. I won the typical Pyrrhic victory that the system allows from time to time. Perhaps I could have gained this minimal success in a fraction of the time by simply greasing a palm or two. But my chance meeting with the failed face of governmental policy across this country, the people who bear the economic and emotional burden of environmental and animal compliance (which are in excess of anything defensible under the freedoms once guaranteed in this country), have made me a lifelong combatant.

Q6. GROUPS & ORGANIZATIONS

The Alliance for America is a coalition of groups who seek a more commonsense, human-centered relationship with resources, and a restoration of the constitutional environment of property rights. In the Alliance I have met sustainable-use proponents who are involved in animal issues around the world. It is the largest organization, both in numbers and in breadth of interests, that I have found for fostering discussion of and promoting sustainable use.

Q7. "THE OPPOSITION"

Any group that advances collective rights, of animals or humans, over individual human rights, is the opposition. Of course, this is akin to tilting at a whole field full of windmills, but it is the only intellectually honest way I see of approaching the problem. While focused on opposing environmental and animal rights interests, one cannot help but be drawn into the broader question of the entirety of rights guaranteed by our civilization.

Q8. HEROES

E.O. Wilson is kind of an anti-hero to me: one whose fame and talent has been turned to "the dark of the side of the force," to use a Star Wars analogy.

I am most inspired by Ayn Rand because of her personal devotion to freedom and human industry. She understood the gift of human consciousness and the innate freedoms it bestows upon its recipients better than even the founding fathers. Or perhaps she understands it in the modern industrial context, whereas the founding fathers posited these noble ideas in an agricultural society.

Rand, for years, has been a trite idiom viewed as being on the extreme fringes of freedom, and she has not really been taken seriously in contemporary American life. Her novels were imagined to be caricatures of the American industrial experience, which should give way to more sensitive philosophies. Now that the lights are going out in California, America can really be seen accurately through the prism of a Rand novel. Her work looks prescient rather than foolish or nostalgic.

Q9. RELIGION/SPIRITUALITY

As far as religion and spirituality goes, Ayn Rand, who is a personal hero of mine, rejected that path. I, however, am more an agnostic than an atheist. I don't believe that religion has a necessary place in public life, but neither do I believe that religion should be banished from the public square, as it has been. I work frequently with people of sincere religious faith and I do not find my outlook to be incompatible with theirs simply because our spiritual foundation is not identical. As I suggested, animal and environmental rights philosophies are nothing but Earth worship. In this context, I celebrate anybody's right to worship the Earth, and even to proselytize. I do not, however, support their right to force these religious convictions upon myself or other Americans.

Faith-based participation in public life is a long tradition. I wouldn't seek to end its place in public speech, but it must be constrained where it would utilize government to abrogate individual freedoms in the name of religion (whether it is a traditional or non-traditional religion). If anything, religion itself has as much to fear from participation in public processes as public processes have to fear from religion. It is an area where caution and definition are important. By defining traditional and fundamental religious elements as unfit for political life, but allowing non-traditional worship into the public square in unprecedented quantity, we have not provided a reasonable balance for public discourse on religion.

In the religious environment, a hero to look to is Lord Acton, famous for his pronouncement that "absolute power corrupts absolutely." He was a champion of limited government and individual freedoms, while also a man of devout faith and position in the church. The two philosophies are not mutually exclusive. I happily associate myself with his tradition, though I am agnostic.

Q10. FUNNY/STRANGE EXPERIENCES

The biggest demonstrations we ever managed were against our happily vanquished vice president who came to Rhode Island to address my environmentalist enemies. I remember standing in the cold spring rain with our 50-ft. banner, emblazoned with "Stop Gore's Genocide of Rural Culture," and debating John DeVillars, the region one EPA director who sallied forth to defend Gore from our depredations. At one point I suggested that it wasn't just nuts like us who stand on street corners with sandwich boards who think that environmental philosophy has gone around the bend. Indeed, I was in possession of a copy of a letter from the governor of Maine accusing the EPA of conspiring against the state with environmentalist opponents of dam relicensing (a suddenly timely topic, as dams in Oregon are being tapped to bail out California from its self-inflicted energy shortage). DeVillars offhandedly responded, "Oh, you mean a nut like Angus King." Well, as luck would have it, a contingent had come to our demonstration from Maine and was only too happy to take news of this slight back to the governor, and it would appear that in consequence of this networking DeVillars ultimately had to eat a little crow. Little victories like this make seeking redress of one's grievances just a little less tedious and a little more fun.

Q11. LESSONS/ADVICE

[No response.]

Q12. ADDITIONAL INFORMATION

[No response.]

Robert Cohen

Q1. BIOGRAPHICAL PROFILE

I am Robert Cohen. I used to do research in the field of psycho-neuroendocrinology. I did animal research for many years. In 1994, almost eight years ago, I read an article that was in the world's largest newsletter, *Health and Healing*, discussing the greatest controversy in the history of the Food and Drug Administration, the approval process for Monsanto's genetically engineered Bovine Growth hormone. The author was Dr. Heimlich's wife, Jane Heimlich, who subsequently wrote the Foreword to my book, *Milk: The Deadly Poison*.

Upon reading this, I wanted to find out, as any scientist would, if that genetically engineered milk was healthy. I have three daughters, and I wanted to see if it was good for them. I was probably the biggest ice cream and cheese eater in the country; I had nothing against milk at that time. So I requested a lot of scientific data, which being an ex-scientist I was able to read, extrapolate the information, and interpret the data. That is what I did, and I saw a lot of problems.

Now I see my mission literally as hunting the white whale, or tilting at windmills. Milk has become my major issue.

Q2. BECOMING INVOLVED

It was a gradual process. The persons having the greatest influence on me were Howard Lyman, "the mad cowboy," and Neal Barnard, the Director at the Physicians Committee for Responsible Medicine. I read books. One book that really did it for me was Gail Eisnitz's *Slaughter-*

house. I have an insatiable appetite for reading; I average about two books per day. Animal rights books are the books I wanted to read; they have been a fascinating eye-opener.

This has been a long journey for me. I used to eat meat and even hunt. I did all the things that I am now against. For years I was a researcher; now I do not believe in animal research in any way, shape or form. I think it is a total betrayal to animals. But it was a learning process for me. I went from eating a standard American diet, to giving up all milk and dairy products. That took me a couple of years to accomplish completely, because I first found an organic source of milk and dairy . . . only later did I realize that it wasn't just the genetic engineering that was the problem, it was all milk. Then I learned about animal rights and animal abuse, and decided not to eat their bodies any longer. That is where I am today.

Q3. IMPORTANT ISSUES

Seven days a week, eighteen hours a day, I spend on milk and dairy issues. I believe that every sip of cow's milk contains virus, pus, bacteria, powerful growth hormones, and proteins that cause allergies, antibiotics, pesticides, fat, cholesterol, and dioxins. My mission is to empower physicians to get together and express what they know deep in their hearts, that cow's milk does not do the body any good. From A to Z, from allergies to zits, I have shown that milk and dairy products are very dangerous for the human body.

Instinctively, we know not to drink human breast milk. We get sick at the thought of drinking dog milk and pig milk and horse milk, because we know these substances have lactoferrins, and immunoglobulins, and hormones. That is what milk is now, and what it has always been: a hormonal delivery system. The average American every day, eating his milk and dairy, eats the same cholesterol contained in 53 slices of bacon. That is a heart-clogger! We know that animal fat and cholesterol is dangerous: heart disease is America's number one killer.

I made a discovery about milk. There are 4,700 mammals in the animal kingdom; hundreds of millions of different hormones in nature. Only one hormone is exactly alike between all these species. That hormone, discovered only twenty years ago, is called Insulin-like Growth Factor, or IGF-1. That hormone has been called a key factor in the growth and proliferation in human breast cancer, prostate cancer, and lung cancer—and it is identical in cows and humans.

Q4. GOALS & STRATEGIES

The dairy industry recently wrote an editorial about me in their magazine called *Hoard's Dairyman*. They wrote that as a result of my work, the dairy industry now has forty people working full-time to try to counter what I am doing. My goal is to have four hundred of them working full-time to counter what I am doing. I get a lot of media attention. I write a lot of articles.

I write a column every single day. It takes a lot of dedication; it is hard work to write a column every day. But in doing so, I have made a lot of discoveries. I have discovered many of the diseases in milk. I work with many physicians and leading-edge researchers. I am working with doctors on Crohn's Disease. One hundred percent of people with Crohn's Disease test positive for a bacteria that is passed on in the milk. I work with people on cancer, in cancer studies, heart disease studies, and diabetes studies. I learned that there is a protein in milk that is a trigger for Insulin Dependent Diabetes Mellitus. I have learned that early exposures to these proteins, particularly casein, can cause children to die in their sleep because their bronchials fill with mucus. Casein is eighty percent of milk protein, and is the same protein that is the glue that holds a label to a bottle of beer.

Recently on my website, *www.notmilk.com*, I have written a list of A to Z, from allergies to zits. You give me a letter of the alphabet, and I will give you at least a half dozen scientific studies in peer reviewed journals that will give you a link to milk. And I am not the only one saying it; I have quoted some of the greatest physicians in history. I have taken Dr. Spock, who in the last version of his book said that no child, no human, should ever drink cow's milk. We talk about iron deficiency, and the bacteria in milk, lactose intolerance, pesticides, nasal congestion, and arthritis. By eliminating milk and dairy, it is like the answer to "the fountain of youth."

We haven't even mentioned the animal abuse, what happens to these cows. One of the greatest effects that I have had on this movement is that I have gotten vegetarians to become vegans, people who do not eat milk and dairy products. The most horrible form of animal abuse is what we do to cows on dairy farms. A pig or beef cow is quickly led to slaughter. These creatures are kept in horrible confinement, and they are slaughtered. A cow, in modern times, has an average life of only 35 months. In fact, people in slaughterhouses can tell which cows were treated with the genetically engineered Bovine Growth Hormone because they can break the pelvic bones with their hands. Imagine these diseased animals going into the slaughterhouse. Thirty-five months! Cows used to live on the farm for twenty-five years! Now they live just three years. They can hardly even walk, they have such tremendous, horrible pain. I recently wrote a column called "What Happens to the Udders?" We are eating body parts from diseased animals in the name of good health. The USDA has recently changed its standards to allow cancerous animals to be slaughtered.

It is a horror story, what happens on the dairy farm. The female cow has to be pregnant to give milk. When a cow gives birth, that cow is separated immediately from the calf, on that first day. The crying on that farm, of mother and infant, causes nightmares for the human children on the farm. That is why many children do not follow through and become dairy farmers themselves. It is a tortuous existence for these animals. The male calves become veal. There is a great movement, which I am very happy about, to ban veal, even in the finest restaurants. This is one thing that Farm Sanctuary is doing, and they are very successful at it, in New York City. These animals are taken from the farm and shoved into a truck. They have tremendous fear, because these are very loving animals. They are driven to the slaughterhouse, and some of them freeze to death in the trucks. They are dragged into the slaughterhouse, and the stun gun isn't powerful enough to completely knock the animal out, because if they put it on full stun, it will burn the flesh and destroy some of the meat. So they are just stunned enough to tie a rope around their leg to lift them up. As they are kicking, somebody takes a knife and sticks it in their neck. The blood gushes out: not just a little bit, but gallons and gallons. Sometimes these animals are still alive, mooing, and shaking spasmodically. They seek any way to escape, but they are pulled down the slaughterhouse line kicking, with their blood mixing in with their feces. They are choking in their blood, eyes bulging. They kick out, sometimes injuring the people working in the slaughterhouses. It is a horrible thing to see. And to think about eating the flesh of these animals: eating the fear, the adrenaline. If you want to talk about animal abuse, look at the dairy farm.

Q5. DEFINING SUCCESS

My ultimate goal is to have everyone in America repeating my mantra, "Pus with hormones and glue." The Food and Drug Administration in America allows every liter (about a quart) of milk to contain 750 million pus cells. Every sip of milk has estrogen and progesterin and testosterone and prolactin and melatonin, 59 powerful bioactive hormones. Pus with hormones and glue. Eighty percent of milk protein has this substance casein, which is the glue used to hold together the wood in furniture. You eat casein in milk, cheese, ice cream, and even tuna fish (they put casein in tuna fish to hold it together, read the label)! They put it in Hostess Twinkies, and in Aunt Jemima Pancakes. The only cereal on the shelf that has it is Special K, which is what makes it special. This is a tenacious glue. You eat it, and your body produces histamines. You make histamines, and you make mucus. It doesn't happen in ten minutes, it happens about twelve hours later.

I want everybody in America to take a test. Go one week without eating any milk, cheese, butter, yogurt, sour cream, cottage cheese, whatever: no dairy products for one week. On day eight, just eat one slice of pizza and see what happens ten or twelve hours later.

Do you know that the average cow in America is infected with one of five different bacteria? Mycrobacterium paratuberculosis causes Irritable Bowel Syndrome; thirty million women in America are affected by it. Salmonella, E coli listeria. Mycobacterium Para-T, these are powerful bacteria. We constantly see recalls from the Food and Drug Administration for Listeria, or E coli 157, or Salmonella. The Center for Disease Control said that there were 175 million cases of food-borne illness in America last year.

You have milk. Do you think pasteurization works? I want Americans to know that the milk in your refrigerator, which you smell on day ten, and pour it down the drain . . . what do you think you were drinking on day nine? That is what we are talking about. At the first sign of pasteurization, these rod shaped bacteria form a spore within their cell. Spore is a Greek word for seed. When the milk cools, the seed re-emerges to its form and starts growing. It grows in the refrigerator, at forty degrees Fahrenheit, and takes about thirty hours to double. At room temperature it doubles in twenty minutes. That pus in your milk, the 750,000,000 pus cells, and the 20,000 live toxic coliform bacteria (that the FDA allows) in every liter of milk, grow and multiply. It takes Listeria forty-five days to grow in your body and make you sick. You are not going to remember eating that cheese today and then getting sick forty-five days later. That is what milk does to you.

Q6. GROUPS & ORGANIZATIONS

I am not a member of an organization other than my own: I am the Executive Director of the Dairy Education Board.

I support and admire many animal rights organizations and people. I admire Farm Sanctuary, and the People for the Ethical Treatment of Animals, and the Physicians Committee for Responsible Medicine, and Voice for a Viable Future (Howard Lyman's organization). I attend many animal rights conferences, and I speak and lecture. I am very proud to be a part of this movement.

Q7. "THE OPPOSITION"

I don't see any opposition, I just see a lot of ignorance out there. I think, in theory, that every human has humanity. Every human, if given the right amount of truth, will come to realize exactly what we are trying to accomplish here. We are trying to have you look at an animal, and look into its eyes. Look at the pet you have, the dog or the cat that you fell in love with. This animal has feelings too. So does a cow or a chicken. It is painful to introduce sharp pointy things into your neck, especially cutting into your jugular vein or carotid artery. It is a painful experience to die.

There are no opposition groups, only ignorance. I think of Albert Schweitzer, the great humanitarian. A week before he died, he wrote a very famous letter. At that time he was not a vegetarian. It hit him as he was sailing up a river and saw a hippopotamus yawning, and he said, "that's it, in humanity we are all the same spirit, the same soul." We may look a little different, we all have love, and form emotions, and form cognitive thoughts. We all have that same bond; we do not want to die. We like to eat, make love, and have children. We enjoy the same emotions. We see this over and over again in the animal world: they form family bonds and emotional bonds. They form conceptualized thoughts, they plan; there is very little difference. If you respect the rights of humans, your own family, and your own dog, you have to take it to the next level.

There are no adversaries, only ignorant people who one day will become one with us. I was an animal researcher, and a hunter and fisher, but I learned the truth.

Q8. HEROES

Howard Lyman inspires me the most. Howard Lyman is the "mad cowboy." Because of him, I am a vegan. He is tireless, out there working every single day. Give him a phone booth with one person, and he will be there, explaining and lecturing.

Another one is Gail Davis, who runs an animal sanctuary for chickens, who is a great inspiration. These are some of the people in the movement who waste no effort reaching one or a thousand. It is a seven-day occupation for them, they have been touched by it. These are people who are brilliant, talented, and innovative, who could go out and make millions of dollars in any field they choose, but they choose to make a difference on this Earth.

Q9. RELIGION/SPIRITUALITY

In a sense, religion means to me, oneness with the universe. Oneness with a universal wisdom that we should all be connected to. This is something that anyone in the animal rights movement feels, that universal wisdom that makes the dog and the cat and the pig and the cow and the chicken, all the same creature. Whether one is a Jew or Christian or Muslim or Buddhist, we all come from the same spirit and have the same feelings. Look into the eyes of your pet dog or cat. Stroke the cat's thigh and look at it: it is very much like the chicken thigh that you are about to eat! Think about that next time: think about your cat.

I call this universal wisdom. Whether one believes in God or Mother Nature or evolution, one must respect the great wisdom that brought us to where we are now. We are very complicated, humans. We have three trillion cells. And every one of those cells has thousands of electrical and chemical processes every second. I marvel at the complexity of this system, of things that we have not come close to deciphering and understanding. But the one thing we do understand is that each one of these cells has pain receptors, whether in a dog or cat or human. Just as we do not eat other humans, or accept harming other humans, I feel that putting a knife against an animal is the same act as putting a knife against a human. People ask me about this. Can we justify killing anything? No, not even a moth. I have been taught that whenever a moth comes (causing some conflict at home!) I let it go. I do not kill moths or mosquitoes. Every creature has a right to live.

Q10. FUNNY/STRANGE EXPERIENCES

When I first started in the animal rights movement, I didn't know what to do. I didn't know anyone in the movement. I just knew that I saw a problem with milk. I decided to go to a supermarket to pass out flyers. So I went to a supermarket in Teaneck, New Jersey. I took an armful of flyers and started handing them out and warning customers about the dangers of genetically engineered milk.

Twenty minutes later, a police car pulled up with the lights on. Out stepped two policemen, putting their hands on their belts, and they looked like they were about to tell me that I was in big trouble.

One policeman told me that I would have to leave. I said, no, I am not going to leave. He said, "If you don't leave I will have to arrest you, because you are on private property."

I said, "Excuse me, but this isn't private property." He said, "Yes, you are standing in the supermarket's property; they own this parking lot."

I said, "They do not own this, they just rent it." He said, "They are the tenant and they have complained."

I said, "Well, I am the landlord." So I pulled out the lease and proved that I owned the property. I knew I was not going to get into any trouble. The supermarket was my tenant. I was a real estate developer.

Q11. LESSONS/ADVICE

You must persevere. You will often be disappointed. You will have ninety-nine disappointments for every one good experience. You must try to average out those experiences to see that you are getting somewhere and making a difference.

My religion and the Jewish religion say that if you help one person, you help the world. I get over 2,000 e-mails a day on my website. It gets to be a burden sometimes. That is why being a vegan is a good thing—I only need four hours of sleep every night. As a vegan I have more energy and can spend more time working.

I have cured a number of people with the most unusual problems, ailments, and diseases. Children with diabetes, children with brain diseases or asthma—we have been able to turn them around. I like making a difference.

Are we winning? Damn right! Five years ago I couldn't go into a supermarket and find soy milk. Today every supermarket in America has a special section for soy milk. Soy ice cream. Soy veggie dogs, soy hamburgers. We are winning; people are becoming vegetarians. Just yesterday I saw that Heinz has come out with a new line of soy products. Now they are a little bit ignorant: one of their new soy products lists a recipe containing yogurt. I called their marketing people to ask why they are manufacturing soy and promoting yogurt. So we have a little work to do, teaching them. But we are seeing more people becoming vegetarians.

I am very much committed to animal rights. But I started in this movement because I was primarily concerned with compassion to humans. That is what this is all about. By eating animals we show no compassion to our own bodies. We hurt the animals, of course, but this is also about human rights. The average American is eating three vegetables: french fries, pickles, and ketchup. You go into the supermarket, and there are forty vegetables and forty fruits to choose from. There are twenty different grains and twenty beans. There is a tremendous variety. There is no reason in this day and age to eat animals for nourishment. There is no reason to kill them, torture them, cut their body pieces into little parts and bread them and flour them and fry them, when the soy equivalents taste just like the meats. The fruits and vegetables are so fresh and delicious!

I live near New York City, which has some of the world's finest restaurants. I go to Gallagher's Steak House for a business meeting. Everyone else orders steaks. I ask the maitre'd to bring me the chef. I order the portabello mushrooms, the polenta, and the garlic mashed potatoes, and the broccoli rabe, give me ten different vegetables, your finest side orders. When it is delivered, everyone else is jealous. We have food out there that has much more flavor and texture than animal flesh.

Q12. ADDITIONAL INFORMATION

As an animal rights person and ex-researcher, I can tell you that there is no longer any need for animal research. It is a complete betrayal. With human tissue samples being stored in labs, with computer models, animal research does nothing. I challenge people to debate: teach me one thing that we have learned from animal research that can help humans.

Humans are not rats. Rats do not have gall bladders. Half of the cancers that rats get, mice do not get. We cannot take that information from a rat and apply it to a human. Nor can we take it from a chimpanzee and apply it to a human. The drugs that have been approved, like Thalidomide, did not produce any adverse reactions in animals, but produced terrible effects in humans.

Vivisection only promotes the researcher's grant money and career; it does nothing for the animal or the humans who rely on the research.

An essay written by Tolstoy, called "The First Step" [essay, 1893, read excerpts at *http://ivu.org/history/tolstoy/step.html*], inspires me. In the essay, Tolstoy makes a philosophical and intellectual argument about the first step we would take in becoming vegetarian. He takes us through a Russian slaughterhouse. He makes a good argument that we have developed in our evolution. We no longer need to eat animals, and we can no longer justify it intellectually.

Priscilla Cohn

Q1. BIOGRAPHICAL PROFILE

I was born in Philadelphia in 1933, one of five children. I am a widow, a mother, and a grandmother. I am also a professor of Philosophy at Abington College, Penn State University. I am a graduate of Bryn Mawr College having received my B.A. cum laude with honors in Philosophy in 1960, M.A. in 1962, and Ph.D. with a dissertation on Heidegger in 1969. I have published articles and books on a number of philosophical topics in both English and Spanish and have given papers on five continents: North America, South America, Europe, Australia, and Africa.

I enjoy reading both fiction and nonfiction, music, theater, movies, and dance. I am entranced by the moonlight on newly fallen snow. The dappled sunlight through green leaves delights me. I feel privileged to glimpse a tiny mouse, almost hidden by grass, and to hear the sound of his eating I know not what. I love a horse's soft nose, a cat's velvet paw, the haunting cry of zebras as they line up to drink and then thunder off frightened by something unseen. The lion's roar fills me with wonder. I am charmed by elephants surrounding a tiny baby to protect it and fascinated by the way a female bird will drag her wing to lure danger away from her fluffy chicks. The glimpse of silver fish darting through clear waters thrills me, and the grace of seals as they twist and turn in the liquid, alien pale green world below the ice enchants me. I am bewitched by the fragile beauty and amazing strength of the butterfly, the tiny legs of an insect as it crawls up the stem of a flower. The natural world fills me with awe . . . and fear for its well-being.

Q2. BECOMING INVOLVED

I grew up with animals. My family had English Bull Terriers in the house, English Setters and cats in the barn, cows, horses, a pony and a donkey, a pair of mules, chickens, guinea fowl, and turkeys. My brother had rabbits for a while. We all knew that there were snapping turtles in the pond, raccoons, groundhogs, skunks, and even foxes in the woods and meadows. I learned that animals are individuals and that each has his own personality. As I became intellectually aware, I realized that loving animals and respecting them are not the same thing. More important than responding emotionally to animals, I think, is respecting them, which means that they ought to be able to live their lives free from human exploitation. That sentient beings are due respect follows from rational argument, and my respect for rational argument was intensified by my study of philosophy. To answer the question, my involvement in animal rights was a gradual process in which I became aware of animal suffering in food production, in laboratories, and so forth. Still, I have always been fond of animals; even as a tiny child, I snuggled up to the dog or tried to sit on the donkey as she was eating grass in the field. One of my greatest joys was feeding the horses on a chilly winter evening, feeling the warmth of the barn, savoring the odor of horses, grain and hay, hearing their chewing, the sound as they moved in their stalls, and the exaggerated ticking of the large clock that never told the correct time but never stopped working.

Reading about intensive farming conditions; the many repeated, needless, and senseless experiments performed on animals; the practices of the fur industry; and so forth persuaded me that at least in the classroom I must stand up for these most vulnerable of living beings, these innocent creatures. I was certainly informed of many of these ills by reading Peter Singer's *Animal Liberation* in the seventies, but my dislike of violence was undoubtedly influenced by my early education in a Friends' school (Quakers). A deer hunt in a relatively small park to which most local people objected made me aware of our war on almost all forms of wildlife and was the impetus for me to organize a protest and bring a lawsuit to stop the hunt.

Q3. IMPORTANT ISSUES

It would be difficult to try to point to a particular form of animal abuse as worse than another. One would have to decide what criterion would allow one to say that one practice is worse or more horrendous.

Are we talking about numbers killed, suffering endured, or what? I have spent most of the time working on issues concerning wildlife simply because there seemed to be a number of people concerned with the plight of companion animals, food animals, and laboratory animals while wildlife is almost invisible and therefore often ignored.

From a humanocentric point of view, wildlife is important because many people enjoy "nature." That is not my point of departure. I believe that almost any conception of justice will show that we cannot justify killing wild animals for their fur, feathers, skins, and so on, simply because we think that it is pretty; nor can we chase, terrify, and kill animals simply because doing so allows us to experience our power. Recreational hunting is killing for fun, and I am opposed to killing in all forms, unless it is clearly a rationally established matter of self-defense.

Most people, I believe, desire to be treated justly. Consider all the hot spots in the world today and you see that the complaint is that people are treated unfairly. Most people understand what justice is, and most agree that inflicting pain and suffering on another living, sentient being without some important purpose is improper from a moral point of view. Thus, all that remains to be shown, in issues such as hunting, is whether the recreation is what we would call an important purpose.

In sum, I care about these practices because I think they involve important moral issues. I think other people should care about them because a world in which might makes right is a brutal, violent world full of pain and suffering for all involved, human as well as nonhuman animals. Similarly, more compassion, more empathy can only improve the world for all involved.

Q4. GOALS & STRATEGIES

I do all of what I've discussed. I believe the public has the right to know the truth about the transportation and slaughter of food animals, the treatment of animals in laboratories and the conclusions that can or cannot be drawn from experiments, the ceaseless manipulation of some animals to provide targets for hunters, and so forth. I believe it is important to teach people to question the sometimes superficial explanations that are given for animal abuse. In sum, both my short- and long-term strategies involve education, "letting the light in," revealing what is behind the closed door. Sometimes this door is literally a door, but other times it is simply practices and habits that are so customary or habitual that we do not question them. Many people who do not object to children earning spending money by trapping do not realize that it often means stamping animals to death or letting them drown.

Q5. DEFINING SUCCESS

My goal is to approach the vision described in the following words:

> "Then the wolf shall live with the sheep,
> and the leopard lie down with the kid; . . .
> the infant shall play over the hole of the cobra,
> and the young child dance over the viper's nest
> They shall not hurt or destroy all in my holy mountain." (Isaiah 11:6–9)

These lines express an ideal, not a goal, but any advance toward achieving a peaceful world is a form of partial success. Instead of talking about humans and animals if we think about human animals and nonhuman animals, we will, I believe, come to respect the other inhabitants of this planet, and we will appreciate differences rather than taking a hierarchical view where we consider ourselves superior to all other creatures. Such a view, I believe, would not only allow us to see moral issues more clearly but also is in keeping with Darwin's insights. I would say that we have succeeded once the equation, and its implications, that 'might equals right' is rejected by thinking people.

Q6. GROUPS & ORGANIZATIONS

I am director of Pity Not Cruelty (PNC) Inc., a small, nonprofit organization that has organized conferences and funded research on animal contraception, aided the identification and count of amphibians, helped pay large veterinary bills, and so on. Over the years, I have worked with a number of national organizations such as Fund for Animals, People for the Ethical Treatment of Animals (PETA), and the Humane Society of the United States (HSUS). I was a board member of the Fund for Animals for a number of years, of a local shelter, and of a wildlife group.

Q7. "THE OPPOSITION"

I believe that such a question distorts the issue. I am not against any person in particular, or any group in general, although I certainly believe that certain groups encourage behavior that is improper, which is not to say that everything any group espouses is improper. For example, I am opposed to docking dogs' tails, but I am certainly not opposed to veterinarians. I oppose hunting, but I am not opposed to target practice, and so forth. I am opposed to certain kinds of behavior, not to groups or types of people. I believe that considering certain people or groups as the opposition is too simplistic, and encourages us to think in terms of stereotypes. There are not too many of us who walk on water.

Q8. HEROES

I have been touched by a number of stories: the story about St. Francis saving a wolf, the story of Leibniz replacing an insect on a leaf, the story of Abelard and Thibault and the dying rabbit. I have been impressed by the words of many thinkers, such as Schweitzer when he said, "I am a will to live, surrounded by wills to live." I have been inspired by the courage of Vicky Moore, who did not let considerations of her personal safety interfere with her desire to halt so-called "fiestas," where for no reason at all animals are tortured in small villages in Spain. Years ago I was touched by the friendship and tenderness of Luke Dommer.

Q9. RELIGION/SPIRITUALITY

Organized or institutionalized religion plays little part in my life. On the other hand, moral concerns play a large part in my life, for I believe that we ought to try to make the world a better place. People of privilege, in particular, have an obligation to improve the plight of the less fortunate, be they human or nonhuman animals. I also believe that the ruthless treatment of animals, for fun or for profit, coarsens or destroys the highest aspect of our humanity.

Q10. FUNNY/STRANGE EXPERIENCES

Most of my experiences as an activist are sad or frustrating rather than funny. Let me relate a funny, but revealing story about an animal I "rescued." Shortly before Easter eleven years ago, I bought a tiny goat who was destined for the dinner table. Because she was "meat on the hoof," I paid for each of her fifteen pounds. Although I planned to place her on a farm, Penelope Goat became part of the family. I quickly realized that Penelope was a very intelligent animal, learning words like "cookie" almost immediately, much faster than my dogs. As her horns grew, she knew exactly where they were: She could turn her head so that she could reach into a narrow space. It was puzzling, therefore, that she seemed to get her head stuck between the planks of her stall. I would find her with her head drooping, looking sad and patiently waiting for me to push and pull and finally free her. Several days later, she would be stuck again. I related these incidents to a friend, remarking that this did not seem to be very intelligent behavior, and yet I knew that Penelope was quite intelligent and learned quickly. On questioning me about how she behaved when she was stuck and how I behaved when she was stuck, the situation became clear. Shortly thereafter, I again found Penelope stuck, her head hanging down pathetically. In a firm voice I told her that I knew she was not stuck and that she only wanted the potato chips that I used to give her to try to urge her to move her head, and that I was not going to struggle with her anymore. I then walked away. When I turned around, Penelope had freed herself, and she never got "stuck" again.

Q11. LESSONS/ADVICE

I believe that activism must be based on a clear understanding of, and familiarity with, the issues and the facts. In addition, activism requires practical knowledge. Because our civilization is based on animal pain and suffering, I think it is most efficient if one chooses to focus on a limited number of issues. It is important to realize, I think, that different people have different skills and different talents to offer. Finally it is important to understand that progress is not linear; that we are speaking for the most vulnerable beings; that if we cease our efforts, they are doomed; and that we cannot be in the wrong if we are opposed to death, pain, and suffering.

Q12. ADDITIONAL INFORMATION

[No response.]

Karen Coyne

Q1. BIOGRAPHICAL PROFILE

My name is Karen Coyne. I have been vegan for nearly a decade because I believe the most important way to respect the rights of animals is to not kill or eat them. California has been my home since I was two weeks old, but I was born in Indiana. My grandparents were farmers back in Indiana and actually raised cows to be butchered. I remember as a child thinking that that seemed kind of sad, but it's just the way things were, and I figured adults knew what was right. Then I grew up. Now I enjoy a cruelty-free diet, and I've learned to think for myself and listen to my conscience, rather than other people, as a voice to guide me to do what is right. I have been an entrepreneur for the past four years and run a successful home-based business. However, I'm ready for a career change, so I plan to start teaching high school English in the fall. I enjoy a number of things, such as spending time at the beach with my dog, doing outdoor activities, and going to the movies. I think it's important to enjoy life, and I'm thankful for the freedom to do so. But I believe that all beings should have the right to those same freedoms, especially the freedom to live a life free from abuse. Until that day comes, none of us can find complete happiness and enjoyment in life, while some of us are still suffering.

Q2. BECOMING INVOLVED

I became involved in animal rights issues soon after I discovered the travesty of factory farming and the power that the meat, dairy, and egg industries have in controlling the "facts" we learn about a proper diet.

I was in my junior year of college in January of 1992 when I met a "vegan," a term with which I was not yet familiar. Being the curious person I am, I wanted to know all about my new friend's diet. He gave me some reading materials written by John Robbins. After a few months I finally read them and have never been the same since. I continued to learn more about the horrors and abuse of animals and how the public has been misled, and I thought everyone would stop eating animals if they just discovered what I had discovered. Then reality hit, and I learned the hardest part of it all—people wouldn't listen, and that's when I knew I had to devote my life to this cause.

Q3. IMPORTANT ISSUES

My activism mainly takes the form of veganism and the protection of factory farm animals. Although all animals share equal importance in my mind and all of them need to be heard through the voices of compassionate people, I focus more on "factory farm" animals (cows, pigs, chickens, etc.) for many reasons. For one, the combined number of animals being killed for fur and those in laboratories and shelters pales in comparison to the 9.8 billion birds and mammals that were bred and killed for food in the United States last year. Although every animal is important, we can prevent the most amount of suffering and make the most significant impact with our forks. I'd like to have the time to help other animals such as those sitting right now in laboratory cages hooked to machines, or those stuck in steel-jaw clamps waiting to be killed for their fur, but activism for these causes *requires* action and time in order to accomplish anything. As a vegan, I know I am making a difference even as I sleep. All I have to do is not eat animals and their parts and fluids, and even if I do nothing else, I know I have made a difference. Unfortunately, because vegetarians make up less than 2 percent of the population (the number depends on whom you ask) and vegans are even much harder to come by, an increase in activism is necessary in order to put an end to animal abuse. As Edmund Burke said, "The only thing necessary for the triumph of evil is for good men to do nothing."

In compiling my reasons why I care so much about animal rights and being vegan, ethical reasons are decidedly at the top of the list, and health is somewhere near the bottom, with innumerable reasons between. I often find that the people who stop eating meat for ethical reasons as opposed to health reasons are the people who stick to it forever. I actually stopped eating "red meat" in junior high after learning it is very high in fat and cholesterol, but because I really didn't know enough about it, I would "cheat" sometimes and eat "red meat." Now that my reasons for being vegan are so clear, I would not take a bite of an animal's body part for all the money in the world. Any time I hear self-proclaimed "vegans" saying they "cheated" and ate dairy, I know they are not "vegan" for the same reason I am. "Cheating" implies that they have been denying themselves something that they really want, whereas for me, anything with an animal product in it should be buried, not eaten.

There are so many reasons why other people should also care about animal rights and veganism. The facts regarding the harmful effects of a meat-based diet can be found in numerous books, pamphlets, and Web sites. Television is not a reliable source for the information because so much of the funding for TV programs is from the meat, dairy, and egg industries. I've often heard people who eat meat say it is their right to eat what they want, and they should be left alone because they are not hurting anyone else. Unfortunately, their diet does adversely affect me (and everyone else) in many ways. One reason is the fact that, as a taxpayer, my money supports government subsidies to livestock farmers, which means part of my hard-earned money is used to support something I vehemently oppose. Additionally, a meat-based diet is extremely harmful to the environment we all share. The most important reason, however, that I am affected by other people consuming meat is the fact that innocent, defenseless animals are being abused and eaten, and I am compelled to be a voice for the voiceless. I'd bet most people could not witness a dog or cat being abused and eaten without interfering in some way. I just happen to realize that *all* animals deserve this same kind of compassion. A popular bumper sticker comes to mind: "If you love animals called pets, why do you eat animals called dinner?"

Q4. GOALS & STRATEGIES

Finding the most effective strategies for achieving my goals with animal rights can be tricky. Although becoming vegan was the absolute best thing I've ever done, it has also been the most challenging in many ways. It's not the actual diet that is difficult; it is the people who are opposed to anything that is not "normal." Being different can be tough, but there is something empowering about knowing that I am living my life according to my own standards, not someone else's. The most frustrating and challenging part about trying to educate people about veganism is that most people refuse to step out of the box of the acceptable norm. My goal is to help people step outside that box. The most important strategy I have used to work toward achieving this goal is to teach by example. I do everything in my power to live my life in accordance with my beliefs. This strategy works better for me than actions such as protests or any form of violence and aggression. Instead, I resort to writing as well as one-on-one education, but only to people who are interested. I try not to preach, but sometimes people perceive activists as preachers because they are introducing something new and different (which can sound like preaching). Eventually I would love to reach a broader audience by writing novels and screenplays. Fiction interests me because I have found that simply stating the facts to someone is often not as effective. I try to target the heart and conscience because until people are emotionally involved, they often will not make an effort to change.

Q5. DEFINING SUCCESS

My ultimate goal on this issue would be to experience a world free from animal cruelty, which to me means that everyone would be vegan. I don't count on ever being able to say "we have succeeded" at this in my lifetime, although anything is possible. History has shown us that major changes can occur in relatively short spans of time. As Mahatma Gandhi said, "A small body of determined spirits fired by an unquenchable faith in their mission can alter the course of history."

Q6. GROUPS & ORGANIZATIONS

I do not work closely with any formal groups or organizations at this time, although in the past I have done quite a bit of volunteer work for various animal rights organizations, all of which shared my philosophy of effecting change by means of peaceful action. One of the more memorable experiences I've had with helping animal rights organizations was as an intern at Farm Sanctuary in Orland, California, where cows, pigs, and other animals were able to live out their natural lives in a peaceful environment. There is certainly power in numbers so groups and organizations can be useful; however, it can often be difficult to achieve a consensus among animal rights activists because we tend to be such free-spirited and individual thinkers. This is an obstacle in the animal rights movement, but certainly one that can conceivably be overcome. There are countless reputable animal rights organizations, animal shelters, and so on that greatly need support in order to continue their altruistic efforts to save the lives of animals, and I very much condone supporting these organizations in whatever manner possible.

Q7. "THE OPPOSITION"

I try not to think of any groups or types of people as being "the opposition." Certainly people who are eating animals and animal products are the target group for whom information on veganism and animal rights needs to be shared. I tend to believe that all of us know in our conscience that killing is wrong, but many of us find reasons not to listen to our conscience and instead find ways to justify our actions. It's amazing how much one learns about human psychology after becoming vegan. People who eat meat often feel uncomfortable around a vegan and look for the support of other meat eaters for reassurance. It's natural for people to want to justify their actions and not feel they are doing something cruel. As long as people who eat animals make up the majority, they will have enough people to support them and protect them from their conscience. An alarming number of people will believe any action is acceptable as long as the majority of people are doing it, including slavery and genocide.

My goal is to help people really think for themselves and avoid the group mentality; therefore, I try not to categorize people or tag anyone as "the opposition." I'd like to help people climb the ladder of moral development toward what Lawrence Kohlberg (1927–1987) called the "postconventional" stage of moral development. Experts on this subject agree that very few people make it to this highest stage of moral development where their informed conscience defines what is right. People at this level act, not from fear, approval, or law, but from their own internalized standards of right or wrong. I think getting to this point is equally important as learning the facts about veganism and animal rights. The small percentage of people who are at or close to this highest stage of moral development will often change to a vegan diet instantly upon learning about it. However, the majority of people never make it to this stage of moral development. They may hear many of the arguments from animal rights activists, but they are unwilling to really listen and change because their actions and beliefs are based on external sources such as laws and peer approval, rather than moral reasoning and their conscience. They often do not need to hear more facts but instead need to know how to deal with what they already know, which often means unlearning more than learning. These are the people I try to target, but I prefer to think of them more as caterpillars awaiting transformation rather than "the opposition."

Q8. HEROES

Even though I have known or read of people with qualities I admire, I cannot say that I have any true heroes. I am inspired, however, by insightful quotes from other people. I have volumes of wonderful, thought-provoking quotes, so it would be difficult to choose my favorite. A few that come to mind are "Live simply, so others may simply live," said Mahatma Gandhi. John Stuart Mill: "It often happens that the universal belief of one age, a belief from which no one was free or could be free without an extraordinary effort of genius or courage, becomes to a subsequent age, so palpable an absurdity that the only difficulty is, to imagine how such an idea could ever have appeared credible," and, "Every great movement must experience three stages: ridicule, discussion, adoption." I like collecting quotes and referring to them because they eloquently represent my beliefs, but I tend to focus more on the idea they represent rather than the person who stated them.

Q9. RELIGION/SPIRITUALITY

Religion and spirituality fascinate me, but I cannot say they have played a significant role in my life, in the traditional sense. I definitely believe that we are what we eat: mind, body, and soul. I have a lot of opinions about religion and spirituality, but I still have a lot to learn in this area. I was not raised in a religious family. In fact, the only time I ever remember going to church as a child was when my family tried out a church in our neighborhood, but the next weekend we read in the newspaper that the pastor was arrested for bombing an abortion clinic. We never went back.

I plan on studying religion extensively at some point in the future. Even though I would like to know more about religion, I'm thankful I was not raised to believe someone else's beliefs. Without an external moderator of my beliefs, I was left to seek answers from within and learned to listen to my conscience. Many people just blindly follow the beliefs of whatever religion has been preached to them the most, but I'm someone who cannot be convinced of anything until I know all the facts. Therefore, I may not ever completely establish a recognized religious belief, but I certainly want to learn much more about all religions, especially because I know how much religion affects morality which affects whether a person believes it is acceptable to kill and eat animals.

George Bernard Shaw said, "We pray on Sundays that we may have light to guide our footsteps on the path we tread; We are sick of war we don't want to fight. And yet we gorge ourselves upon the dead." Everyone knows religions have an endless number of interpretations. Many people actually believe their religion supports eating meat, whereas others in the same religion do not eat meat in part because of their religious interpretation. It is an oxymoron for a meat eater to preach love and compassion. The other night I was up late and happened to see a religious television program that caught my eye. A man was lecturing to a group of adolescents about why people should not spend time on causes such as saving animals and forests. He was mocking people who spend their time on these causes because, as he believed, any cause that was not for Jesus or for people was "crazy" and a waste of time. The impressionable teenagers bought every word this man said. That concerns me because once people believe their actions are morally and religiously justified, and once they are taught not to listen or respect the thoughts of certain groups of people, it can be very difficult to get them to step outside that box and open their hearts and minds.

Thankfully, there is a growing number of people who interpret their religion, whatever it may be, as a doctrine that encourages people to reach the highest level of compassion possible; this means respecting the lives of ALL beings and not killing or eating them. There are numerous quotes directly from the bible that religious vegetarians use to support their beliefs. These include Isaiah 11:6 of the Hebrew Bible, "The wolf also shall dwell with the lamb, and the leopard shall lie down with the kid; and the calf and the young lion and the fatling together; and a little child shall lead them." And Eccles. 3:19, "For that which befalleth the sons of men befalleth beasts . . . yea, they have all one breath, so that man hath no preeminence above a beast: for all is vanity." And in The Koran, "There is no beast on earth, nor fowl that flieth, but the same are a people like unto you, and to God they shall return." There are many more religious quotes and arguments in support of animal rights and veganism than those in support of eating meat. Many religious organizations representing various denominations can explain why any religious justification to eat meat is invalid. I agree with Vern Dollase, who said, "A man's religion is a failure if it has not taught him kindness." And The Lord Bishop of Manchester, DTBC, who said, "My Lords, I once heard it said—and the saying has haunted me ever since—that if animals believed in the devil he would look remarkably like a human being."

Q10. FUNNY/STRANGE EXPERIENCES

Although I cannot recall a funny experience I have had in my work as an activist, one strange experience that comes to mind concerns the time of my arrival for an internship at Farm Sanctuary. When I went up to the barn to see the cows, sheep, lambs, and others who lived there, some of them came up to greet me, and others were much more reserved. I remember the woman working there was calling all the cows by name, which I thought was remarkable because they all looked the same to me. I remember the strange part was that I didn't feel an instant connection to the animals like I thought I would. I had stopped eating animals because I cared about them and did not want to inflict pain upon them, yet when I was face to face with these animals, I didn't feel a very strong connection. As the days went by, however, my connection with them grew to a point where I understood each one of them, and I allowed myself to know who they really were, and their personalities really became clear the more I got to know them. Most of them had been rescued from very abusive situations and were afraid of people. However, in the last weeks I was there, I had made such strong connections with all of them that they would run up to me when they saw me, and they would fight for my attention. I am reminded of this experience when I hear derogatory remarks about animals and that they are "stupid" and do not have personalities, especially the ones that get eaten. What people who make these statements don't understand is that relationships with animals are very similar to relationships with people; it really takes time and openness to truly connect. Even if a person never experiences this kind of bond with an animal, it is so important to at least realize that animals are each unique with their own personalities, and they can feel the pain inflicted upon them. Nothing compares with the love of an animal, but even without this, when people show compassion even to those they don't love, that is when they succeed.

Q11. LESSONS/ADVICE

The advice I would give people just starting to work for animal rights would be to stay true to their hearts and let their conscience be their guide. I could go on forever, but much of what I would say has been said before, so I'll just share some words of others such as Margaret Mead who said, "Never doubt that a small, committed group of people can change the world. Indeed, it's the only thing that ever has." And George Bernard Shaw, "All great truths begin as blasphemies." And Socrates, "My plainness in speech makes them hate me, and what is their hatred but a proof that I am speaking the truth?" And Henry David Thoreau, "There are a thousand hacking at the branches of evil to one who is striking at the root."

Q12. ADDITIONAL INFORMATION

With the knowledge that billions of animals are suffering at this very moment, sometimes people resort to almost anything to make it stop, even violence. No matter how frustrated animal rights activists are, I would advise that they never resort to violence. The outrage and impulses may be difficult to control, but it is necessary to do so in the best interest of all involved, including the animals. Compassion is not something that can be forced. Certainly wars have resulted in significant changes in our history such as putting an end to slavery and the killing of Jews. The problem is, even though laws may have changed as the result of these wars, prejudice and discrimination still exist. Laws only hide cruel acts from plain view; compassion makes them disappear. This holds true especially when dealing with animal rights because even if it were illegal to abuse or kill an animal, this would be virtually impossible to enforce because animals cannot speak out if the law is broken. Although laws may help in some ways, if cruelty to animals is to stop, activists must not use force and violence but instead must serve as guides in helping people find and be led by the voice of compassion in their conscience. I agree with Norman Cousins who said, "Nothing is more powerful than an individual acting out of his conscience, thus helping to bring the collective conscience to life." Just as we are often horrified by the barbarity of all ages except our own, one day people will look back on this place in history and will be shocked at our barbaric treatment of animals.

Diana Dawne

Q1. BIOGRAPHICAL PROFILE

I have been a blind person all of my life. When I was twenty-one, I got my first guide dog. They have been a real gateway in my life. I can't imagine how my life would be without them now. I have two degrees. I managed a medical facility; I was consulting director or surgical consultant, I created the position. I directed for many years. I have two doctorates in Theology and Psychology. The dog was a gateway for me to go to school. I have traveled extensively throughout the United States with my dog. I wrote the book *Venture's Story*, about my guide dog, in which I allowed Venture to tell his own story. I am not a typical person, I really like to get outside and mix it up. A lot of funny things happen to me because I am out a lot.

I love music and studying. I enjoy meeting people and animals.

Q2. BECOMING INVOLVED

I became involved with animal issues when I began using a guide dog. I was dragged into it whether I wanted to or not. It is a process. There are many People for the Ethical Treatment of Animals (PETA) people who really feel that it is wrong to use a dog for any type of service. They feel that we are enslaving the dogs. I have been stopped many times on the street to be lectured about my enslaving of these poor dogs. I would say that in the last ten years, this has become more common. I lived in Palm Springs one summer, where it can become very hot. I was on the bus (I don't drive much!), and the air-conditioning was not working. A lady sitting across from me said "You know, I don't feel a bit sorry for you out in the heat, but I sure pity the dog."

Q3. IMPORTANT ISSUES

One of the issues I spend a lot of time discussing is why it is cruel or not cruel to place such responsibility on an animal. I wrote a book about Venture, a marvelous guide dog. I do a lot of lecturing for children in schools, and a lot of public access things. One reason I feel it is acceptable is that the dogs love it. If you were a dog, and you could be someone's pet dog or a guide dog, you would probably choose to be a guide dog because it is so much fun. Guide dogs get to go everywhere. They can stand right under the signs that say "no dogs allowed," and they have no idea what it means to be left at home.

Q4. GOALS & STRATEGIES

I do all of those things. My long-term goal is education. Writing is a short-term goal. At one time I was doing more writing; hopefully, my computer skills will improve so that I can do some more writing. I am basically articulate, I explain well. I enjoy educating. I like to see the light go on in someone's mind who before could not see what good a dog or a person can do. I also have a goal: many people think that dogs are smarter than we are. Maybe they are right. I know that people do not like to be perceived that way. I do think that dogs think, but their brains are different than ours, and their thoughts are different than our thoughts.

I am a very good handler of guide dogs, but I do not train them. Now I am working with my fifth dog. When a guide dog is working, it is wearing its harness. That harness is kind of like a uniform. Dogs do not have the kind of intelligence that we do, and so when they are wearing their harness they understand that they are not supposed to be petted. The reason for that is that if you are trying to pet the dog while the dog is trying to watch traffic for me, of course the dog wants to be petted, and may compromise my safety.

People feeding my dog really bothers me. Not only does it hurt their strict diets, but people food is not meant for dogs. When the dog gets sick, I have to clean it up. And if they eat much garbage (and people food usually is), they are not paying attention to what I need them for.

Q5. DEFINING SUCCESS

At the end of my life, I think I will have succeeded. It is a challenge right to the very end. Using a dog to better life will continue until my life does not continue anymore. I want people to respect different animals for the wonderful souls they have.

Q6. GROUPS & ORGANIZATIONS

I do work a little bit with guide dog schools, but mainly I work on my own. I am no longer in practice, because I had major health problems. But because of this illness I have had time to do some other kinds of work. My Web site is *www.yumston.com*.

Q7. "THE OPPOSITION"

The opposition is people who are ignorant or uneducated. There is a difference between ignorant and stupid. Stupid people wish to remain that way, ignorant people just don't know. Children are the most wonderful because it is fun to educate them; they like to be entertained. People don't want you to come in and educate them, but people love to be entertained.

Q8. HEROES

I have a lot of respect for Morris Frank, who was the first blind person who used a guide dog in the United States. You can read more about him by looking at *www.seeingeye.org*.

After World War I, they used a lot of German shepherds during the war, and they had all these dogs and all these blinded veterans. Somebody said, why don't we train the dogs to help the veterans? That is how this got started. In time, of course, the training and breeding became much more streamlined, and then they started using other dogs like Retrievers. They have a wonderful temperament.

There are many marvelous organizations that train guide dogs, all over the United States.

I also love Kahlil Gibran, who wrote *The Prophet*. It is very ethereal and spiritual.

Q9. RELIGION/SPIRITUALITY

Yes, religion and spirituality are helpful to my work. Absolutely! I think that the dogs share in spirituality as well. In *Venture's Story*, I allowed a dog to die, so we discussed animal souls. My feeling is that in fact dogs probably do have a soul. There is a part of the dog that exists after it dies, just as there is in us. I don't think that the Bible applies to dogs because to commit sin, the dogs must go against the commands of God, and I don't think dogs have done that. So there is no provision for them because they do not need a provision. They are innocent.

Q10. FUNNY/STRANGE EXPERIENCES

One thing that a guide dog can learn is how to find things. They can learn a telephone; so if you are out somewhere and need to find a pay phone, you can tell them to find one. My dogs have a very large vocabulary for finding things. One important thing that I teach my dogs to find is a mailbox.

The way you do that is to go where you know there is a mailbox. You tell the dog to find a mailbox, then take the dog to the box, and pat the mailbox, and tell the mailbox what a good mailbox it is. So I was there one day patting the mailbox and telling Victor how wonderful he was for finding the mailbox. A dear old lady came up to me, and grabbed my hand and said sweetly, "Honey, that is a mailbox, your dog is over here."

Another time I was in a hurry to go pick up some things, and I was waiting to cross the street at the light. A little old lady came up to me and said, "You know, that dog is just beautiful." I said, "Yes, he is beautiful." And she said, "I have been watching you for several days now and I just don't understand about you." So I asked what she meant. "Well, if I was like you I would commit suicide. It must be terrible going through life as a total reject; a total lemon." I don't know how on earth I did this, but I said, "Being a lemon in life is not so bad. If you check with the furniture company, you will find that the lemon oil brings out the shine. Check with Procter & Gamble and you'll learn that lemon fresh Joy washes dishes right down to the shine. If you do a survey you will find that lemon pie is one of America's favorites. But most important: some of us lemons are just so much fun to squeeze."

The last time I had major surgery I was in the hospital, and my husband brought the dog to see me, because our dogs have visiting privileges. The nurse kept saying to my husband, "We usually don't let dogs into the hospital." He told her to take it up with me. So she came in to me and said, "We don't usually let dogs into the hospital." I wasn't in the mood for this because I was dying for a pain shot. But she went on. "I know this is a special dog, and that the charts say we have to let him in. But I have to tell you, I don't mean to be rude or anything, but your dog makes me so uncomfortable." So I asked why. She said, "When your husband drove up and parked beside me, and here he had this guide dog with him. That really worries me: this man is driving on the freeway using a guide dog?"

Q11. LESSONS/ADVICE

Hang in there. It is fun, and worth every minute of effort. Working with guide dogs helps you meet people, go places, and really completes your life.

We all make mistakes. The most important thing I have learned is that anyone who uses a dog walks in the image of God; it is like having God beside you.

Q12. ADDITIONAL INFORMATION

Guide dogs are very happy. I may go to a seminar, and my dog will lie there quietly. Of course I must make sure he has a chance to go out to the bathroom, and to get water to drink. Venture would tell you (and all my dogs would agree) that if he had his life to live over, he would want to be a guide dog again.

These dogs were bred in a special breeding program, because their gentleness and other important qualities are genetic. When they are eight weeks old, the puppies go to a special home with children, where volunteers love the dogs and teach them to be well mannered around people. Guide dog puppies are allowed (by law) to be taken into grocery stores and other places, because they need to be around people. At about fourteen months, they are returned to the school for further training, and spayed or altered. Some trainers will go into cities, on the subway, with the dogs, because some of us need animals who are accustomed to the city life. A friend of mine arranged to fly with a dog; she exposed it very well to city life. That dog has now been placed with a gentleman in Boston, so that training paid off.

Ryan DeMares

Q1. BIOGRAPHICAL PROFILE

I am a writer and scholar specializing in the human-animal bond as it relates to communication and also as it relates to human and nonhuman consciousness. I hold what is probably the first doctoral degree in interspecies communication. I am especially interested in the transcendent essences of communication, which include oneness and compassion—attributes that are part of the motivating force that drives most animal activists.

 I was born in Milwaukee, Wisconsin, in 1947 and attended high school there. I went on to become a journalist/photographer. My career included a stint for the World Wildlife Fund at its international headquarters in Switzerland. It was in Europe that I was first introduced to the global perspective on wildlife preservation issues and measures. However, I've always loved and identified strongly with animals of all kinds, and from early childhood have retreated into nature as a source of personal renewal. My affinity for the outdoors was at one time expressed through various activities such as underwater photography and rock climbing. One of my passions in recent years, jumping horses, could be considered at odds with the animal rights philosophy. From my perspective, taking a horse well through a challenging course of jumps represents one of the pinnacles of good communication between human and animal, but communication is not the concern of most animal activists. Animal rights people and I also do not see eye-to-eye on swim-with-dolphin activities. Both animal rights activists and field scientists generally take the position that wild animals should be left alone. Yet, from my own research and conversa-

Ryan DeMares. Photo credit: © 1998 Brant Photographers

tions I know that many people who have had encounters with whales and dolphins undergo significant personal transformation as the result of their experience, often to the point of wanting to become involved in animal advocacy on behalf of those species.

Likewise, animal-assisted therapy is frowned upon within the animal rights movement even though it is well known that stroking animals has a mutually calming effect. Taking the possibilities of touch-as-communication even further, some people consider the challenge of touching or stalking a wild animal to be a form of interspecies play, as well as an art that requires spiritual intention. I include such activities in my very broad definition of interspecies communication.

Q2. BECOMING INVOLVED

When I was about eleven years old, I lived with my mother in a house with a big yard. My mother loved animals, so because we had some space, we acquired a lot of cats and dogs. In building this menagerie, she took me to the animal shelter, and we would bring home pairs of kittens. When they died soon afterward, as one or both always did, we would go to the shelter and get two more. Before long, the fourteen-year-old boy who lived upstairs in the neighboring duplex began to boast that he was poisoning the kittens. My mother told me he probably was using strychnine. We would put the kittens in our backyard rabbit hutch at night, so it was easy for him to slip the poison to them. Somehow, no measures were ever taken to thwart his malicious acts.

On one occasion when I found a kitten dead, I noticed that the fur on its head had been licked flat. I immediately concluded that its littermate had been comforting it in its dying agony. I didn't tell anyone about what I had observed, but it became a profound lesson in compassion—the only lesson in compassion that I can recall receiving as a child, because in my family of origin, although I was taught love for animals, compassion does not seem to have been part of the package, which seems odd to me now since love and compassion are so closely related.

In the midst of the kitten tragedies, I arrived at the intuitive realization that my neighbor was likely to grow up to murder people. Today, many years later, the link between domestic violence and animal abuse has been shown, and future studies will undoubtedly reveal more connections.

At the same time that the poisonings were occurring, someone brought antivivisection publications into my home. I still remember the gory pictures. With all of the suffering I had seen already, the pictures didn't shock me, but they did provide a direction for the anger I was feeling inside. I fervently wanted the practice of vivisection to come to an end. In this way, during my formative years, I came to equate the slaughter of my kittens with animal experimentation. Once that period of my life was over, I did not give further thought to animal rights issues for many years, but the seeds had been planted. To this day, although I am deeply sympathetic with all aspects of the animal rights movement, laboratory experimentation is still the issue that I identify with most closely, and the use of poison to control any animal population outrages me.

Q3. IMPORTANT ISSUES

As a scholar, I have been drawn to explore the other extreme of the human psyche—the potential for self-transcendence—in spite of or perhaps as the result of seeing the dark side of humanity so early in life. In my doctoral research, I identified and described the characteristics of human peak experience as it unfolds when triggered by an encounter with a wild animal. I am using the term "peak" in the same sense that the human potentials psychologists have used it: a transformative event characterized by the elated emotions such as joy, exhilaration, ecstasy, and unconditional love. Sociologists have discovered that peak is a relatively common event although some people tend to suppress or deny their experience of it.

Transformation of thought and feeling can also be achieved through a negative experience that becomes a moment of epiphany. The psychologist Abraham Maslow called this the nadir experience. Interestingly, the outcome of a nadir experience often is the same as the outcome of a peak experience in the sense that the aftereffects are positive even if the experience itself was not. I have noticed that people involved in animal advocacy often seem to have turned to it as the result of an insight triggered by either a peak or a nadir experience. What is so pivotal about such an experience is that it becomes the means by which we transcend our ordinary perceptions of ourselves as separate, isolated beings. In the process of a transformative encounter with an animal, we open up to the experience of oneness with that other being. Afterward, our heart begins to open increasingly to all species, our own included. This is a natural consequence of the awakening of our Mind of Compassion, a special kind of intelligence that resides within each of us.

Q4. GOALS & STRATEGIES

My activities and goals combine writing, research, and education. From the time I embarked on the interspecies path, my goal has been to illuminate and promote the human-animal bond, in the hope of achieving a better world for the animals.

At the same time, I hope to continue conducting research. The subject of the wild animal as a trigger of peak experience is rich with possibilities for further study. My initial research inquiry identified the themes and essences underlying spontaneous encounters with dolphins and whales. This particular focus came about because many people report peak experiences in the presence of dolphins. But in my informal conversations I also have spoken with a man who had a peak experience with a moose, and several people who have had a similar type of experience with insects. So, although dolphins definitely are a charismatic species, happily, they do not hold the corner on oneness with humans. Bigger is not necessarily better.

I also would like to develop an accredited educational program in interspecies communication with the aim of providing an academic venue for the increasing number of people who are taking an interest in the subject. It would be offered at the master's degree level and would, at least initially, be modeled largely after my own graduate studies, which is to say that it will focus on nonverbal aspects of communication including bioethics. I do not take a particular interest in imposing human models of symbolic language on other species. To impose neurolinguistics on creatures not equipped by evolution to handle such programming in my opinion accomplishes little more than producing neurotic animals who as individuals are set apart forever from others of their species. Yet language experiments do serve the higher purpose of cultivating compassion in the researchers who work closely with the animals, and they also demonstrate the "human" side of animals—important to establish if our goal for them is personhood. In this way, language experiments are serving an important purpose. As ambassadors for the animal world, the captive Great Apes such as Washoe and the other chimpanzees who are involved in the Ameslan project in Ellensburg, Washington, actually are promoting animal rights.

Q5. DEFINING SUCCESS

We will have succeeded when we have achieved a critical mass of people, each of whom individually identifies with animals to the point that to inflict intentional harm on any other species has become, for that individual, unthinkable. When you feel the existential pain of the animals—your own internal sense of the suffering that is the animals' lot individually and collectively—you literally hurt inside with an ongoing pain. This is an indication that you have in an important sense become one with the Other—not a particularly comfortable level of consciousness to have to walk around in, because of how animals are treated in our society. But it is a necessary awareness for really doing the work on behalf of the animals. And when that degree of sensitivity is achieved individually, we have succeeded, because one more person on this planet has joined the growing collective consciousness on behalf of the animals.

In my more optimistic moments, I believe we as a species can one day in the foreseeable future collectively achieve a compassionate consciousness. Consciousness actually has been identified as the next great area of study in the new millennium, and some experts are predicting that humankind's next evolutionary step will come in the realm of consciousness. So it is not unrealistic to hope for a quantum leap in consciousness that will take us into a new and much more promising era of interspecies relationships.

Q6. GROUPS & ORGANIZATIONS

Although I am not much of a joiner, I have at times worked with or supported various animal rights organizations. Most recently, I have made a long-term commitment to help build The Dolphin Institute (*www.dolphininstitute.org*), a nonprofit organization that is devoted to enhancing the connection between humans and dolphins through education, encounters, and advocacy. I am the director of the institute's Interspecies Communication Division.

Q7. "THE OPPOSITION"

The opposition is larger than a group or type of person. It is our culture. An inherent sense of alienation from the natural world has characterized our society for hundreds of years and in some respects much longer if we consider the biblical ethic of domination of the animals and the earth. This alienation is based on an attitude of separatism and supremacy that, if it continues for even a few decades more, will completely undermine this planet's web of life as we know it.

Q8. HEROES

In the realm of heroics, I especially admire those animal activists who are willing to put their lives on the line for their animal kin. These include such people as Rod Coronado, who recently completed a five-year prison sentence for liberating coyotes from impending death at a federal wildlife control center and then burning the facility down. I don't advocate violence directed toward other people, but Rob and his accomplices did their work at night while the facility was devoid of personnel. This was truly compassionate in action, as well as a self-sacrificing and courageous act.

I also admire and appreciate the caring people on the front lines of animal protection who are doing shelter work and other similar activities. These people must witness animal suffering on a regular basis. Such work requires a special kind of person, someone who is not afraid to experience the pain as well as the joy that can come as part of a sense of oneness.

Q9. RELIGION/SPIRITUALITY

A creation-centered spirituality is integral to my life. Of the major human religions and philosophies, Buddhism holds the most appeal for me because of its emphasis on compassion. Although compassion is a universal teaching of all of the world's major religions, if you look into the rationale behind the Judeo-Christian ethic of compassion toward animals, you will find the cultivation of compassion toward animals is considered desirable because someone who treats animals compassionately is more likely to extend similar consideration to fellow humans. Buddhism, by contrast, does not take a human-centered approach to compassion. In Buddhism, to be treated with compassion is the birthright of all sentient beings. Accordingly, I often find myself turning to Buddhism as the cornerstone of my interspecies philosophy.

Q10. FUNNY/STRANGE EXPERIENCES

One spring several years ago in Seattle, my canine companion LadyBug dug up a nest of newborn field mice. This happened in the midst of the rainy season, and I could not reconstruct the collapsed burrow roof. I felt certain the mice would drown if I left them, yet I hesitated to take them home for special care, knowing that the next day I would be flying to the East Coast to attend a meeting of my doctoral committee. Suddenly, I realized that these tiny creatures could easily travel with me aboard the aircraft.

The next evening, shortly after the plane was airborne, I extricated the mice from their nest in a half-pint food carton and began to feed them with an eyedropper. Soon, I was interrupted by a flight attendant that had spotted my activities. She demanded to know what I had. When I told her they were baby mice, she sternly replied that animals are not allowed in airplane cabins and demanded that I put them back in the carton immediately. She said if I had to feed them, I should do it in the lavatory. The word quickly spread to the rest of the cabin crew, and on my way to the lavatory, I was confronted by another flight attendant who also expressed her annoyance with the situation. I again tried to show the contents of the little carton, thinking that when she saw the blind and very helpless creatures who were less than an inch long, she would see how ridiculous the fuss was. But she recoiled from me. The mice, she insisted, were compromising the integrity of the crew's food service, as well as jeopardizing the health of passengers who might be allergic to them. Then she delivered the ultimate threat: Any animals aboard the aircraft were required to be listed on the captain's manifest. The captain's signature, she said, was required in order to bring anything alive aboard. "You could be fined severely," she concluded.

With that threat ringing in my ears, I fled into the lavatory. When I came out, nothing more was said. Although I was half expecting a reprisal, I wasn't pulled out of the crowd when the plane landed in Chicago or an hour afterward when I boarded my continuing flight to Washington, D.C. So the mice made it to the East Coast. The next day, when I told my mentor about my experience aboard the airplane, he recalled a time he smuggled a litter of coyote pups aboard a commercial flight. In his case, too, the flight crew had been upset. But once they saw the pups, they charmed them. Hearing his story, I gained a new awareness of the subtleties of speciesism. Both coyotes and mice are much-maligned species. But in contrast to the coyotes, the rodents never stood a chance with the flight crew.

Sadly, the adventure ended far better for me than it did for the mice. My attempts at nurturing them were woefully inadequate, and as time passed the core temperatures of those naked little bodies dropped. When I became aware of this, I tried to improve their environment by tucking them in the warmest place I could think of, against my skin, but it was too late. In all, they did manage to survive several days, which was long enough to be present at my meeting in the boardroom of the Humane Society of the United States. Later, when the story got around, I was teased by some of my friends about attending my academic committee meeting with mice on my person, although it was, everyone agreed, in perfect character for an interspecies communicator.

Q11. LESSONS/ADVICE

Attend animal rights conferences. They are extremely useful for networking and broadening your base of knowledge about the issues, which are multitudinous and far-reaching. Just when I think I've heard it all, I learn new things at these meetings.

Also, don't judge others inside or outside of the movement. We all have areas in which we demonstrate a double standard. It is easy to alienate people when we could instead be gaining ground together toward the overall goal. The animal rights movement does not have a particularly good image with many people outside of it who actually are sympathetic to its basic causes, and I can understand why, having been on the receiving end of criticism from activists who did not agree with or understand my choices.

Q12. ADDITIONAL INFORMATION

[No response.]

Sherrill Durbin

Q1. BIOGRAPHICAL PROFILE

I am Sherrill Durbin. I was born in 1956, in Tulsa, Oklahoma, and still live in the area. My nationality could be called "mutt" as I am a Caucasian with little knowledge about my ancestors. My curiosity is overwhelming sometimes, and I could enjoy hundreds of interests if I would take the time for myself. I work full time as a purchasing agent at Tulsa Community College, and the rest of my spare time is spent helping animals through letter writing, phone calling, investigating, protesting, and hands-on rescuing.

Q2. BECOMING INVOLVED

My becoming an animal rights activist was a slow process of evolution and enlightenment through education. I bought a dog at a pet shop at the mall. I didn't know where pet shop dogs usually come from (puppy mills). I didn't know that I was contributing to the torture of mother dogs at these doggie concentration camps by buying Snuggles from the pet shop. Snuggles, however, started my evolution and enlightenment toward animal rights. I saw that she had a unique personality. She had emotions just like ours. It was like having a child, and I realized that if she was that much like us, then all dogs must be. I started volunteering at the SPCA. I found that every dog and cat there had its own unique, wonderful personality, much like humans. I read in the SPCA newsletter about buying products from companies that don't test on animals, and how cruelly animals are treated by companies that do test. Every bit of information I found and read pushed me forward to the realization that all animals deserve our respect and deserve to not be tortured, used, or abused by humans. I acknowledged that all animals feel pain, mentally, emotionally, and physically, similar to humans. I became an animal rights activist.

Q3. IMPORTANT ISSUES

I don't think I spend more time on any one issue than I do others. I know many people do, but I don't. I just can't turn my back on any of the issues. Whether the phone call, letter, or fax is on buffalo, dogs, cats, rats, horses, turtles, or any other animal, it's all the same to me. It must be done, and I do it.

Q4. GOALS & STRATEGIES

My short-term goal is seeing that the judicial systems recognize, prosecute, and make mandatory counseling for people who abuse animals. I want to see slaughterhouses, vivisectors, circuses, and other industries prosecuted for the torture they inflict on animals. My long-term goal is for the majority of people to see that animals are not things; they are living, sentient beings that should not be used and abused. If people understood this concept, I think people would stop eating animals and stop using and abusing them.

Q5. DEFINING SUCCESS

[No response.]

Q6. GROUPS & ORGANIZATIONS

I have worked fairly close with a few groups on various issues. People for the Ethical Treatment of Animals (PETA), United Animal Nations (UAN), and In Defense of Animals (IDA) are some of the groups I have worked with.

Q7. "THE OPPOSITION"

The opposition is industry and people who know the truth about the abuse going on, but continue to do it and do nothing to stop it.

Q8. HEROES

The civil rights activists of the past are my heroes. They risked everything (and sometimes lost everything) to help correct the wrongs being made to the blacks. Many modern-day animal rights activists fall into that same category.

Q9. RELIGION/SPIRITUALITY

Spirituality is a part of my life; religion is not. I feel that all living beings are interconnected spiritually. Even if we weren't, however, I would still be an animal rights activist. Animals feel pain and suffer, just like we do, regardless of whether we are connected spiritually.

Q10. FUNNY/STRANGE EXPERIENCES

A strange experience in my life as an activist was going undercover at a cockfight. I truly felt like an alien on another planet. To see humans cheering and laughing at roosters who plainly didn't want to keep fighting, to see humans watching and betting big money on such a bloody, sadistic "sport" made me ashamed to be a member of the human race and made me question if I really was a part of the same DNA makeup as these people. And for me to keep from crying and running to rescue those poor roosters (whose eyes clearly showed pain and suffering) was one of the hardest things I've ever done. I knew if I did that, I would blow my cover, and also I risked being physically harmed by the violent cockfighters.

Q11. LESSONS/ADVICE

I would tell people just starting out in animal rights activism to try not to take things personally, from animal abusers or from fellow animal rights people. Remember that you are in it for the animals, not for glory for yourself or for any group. Take good care of yourself, because being an animal rights activist is probably the most stressful task you'll ever take on. If you aren't good to yourself, you'll burn out quickly, and then you can't help the animals.

I think a big misconception by "outsiders" is that animal rights activists hate people and don't help people. That is just not true. Almost every animal rights person I know helps people as well as animals. I myself have helped domestic violence victims, the homeless, and any other people I see who need help and cannot or aren't in a position to speak up for themselves.

That is the whole idea of animal rights. We defend and help the helpless and voiceless.

Q12. ADDITIONAL INFORMATION

To end this, I thank my precious Snuggles (Schnauzer/Poodle mix, 3/3/89–11/6/96) for being the one responsible for my journey into animal rights. I love you, Snuggles.

Michael Fox

Q1. BIOGRAPHICAL PROFILE

I was born in England in 1937 and graduated from the Royal Veterinary College, London, in 1962. I immigrated to the United States in late 1962, to begin research in animal behavior and development at the Jackson Research Laboratory in Maine, and then at the Galesburg, Illinois, State Psychiatric Institute. Eventually I moved to Washington University, St. Louis, where I taught animal behavior and became a tenured associate professor in the Department of Psychology. I worked for Washington University from 1967 to 1976, after which time I joined The Humane Society of the United States. For my research into canid behavior and development, I received the Ph.D. (external) from London University, England in 1967, and the D.Sc. in 1976.

I enjoy being with animals, being in wild natural places, vegan cooking, writing, and playing my flutes, drums, and didgeridoo. The didgeridoo is a hollowed-out tree branch or piece of bamboo from 3 to 6 or more feet in length, a traditional musical instrument of the Australian aborigines that has an estimated 20,000-year history.

Q2. BECOMING INVOLVED

My understanding, respect, and love of wolves and the fact that they were being shot in Alaska, and my visits to animal research facilities where I saw the extreme conditions of deprivation for primates, dogs, and other animals, got me involved in animal rights issues in the early 1960s. Decades before, I became aware of how different I was from most other people when I found a bag of drowned kittens in a pond and on another occasion found several frogs that my childhood peers

Michael Fox. Photo credit: Michael W. Fox, Jr.

had blown up with straws. Such acts of cruelty and destruction were inconceivable to me.

Q3. IMPORTANT ISSUES

Currently, I am focusing on human-animal relationships, bioethics, the spirituality of compassion, and the mind-set behind and consequences of genetic engineering biotechnology. I believe it is my challenge and responsibility as a veterinarian to help heal the relationship between humans and other animals.

> Health care = People care + Animal care + Earth care.

People need to realize these connections for their own good, so that they can examine their attitudes, values, and feelings that influence the ways in which they relate to other sentient beings, for better or for worse.

Q4. GOALS & STRATEGIES

I write, lecture, and dedicate all my efforts to these ends, my work being my life. I also believe in putting compassion into action at the grassroots and work with my wife Deanna Krantz, who directs India Project for Animals and Nature, running an animal refuge and free veterinary service in S. India (*www.gcci.org*).

Soon after I joined HSUS as scientific director and head of the Institute for the Study of Animal Problems, with Andrew Rowan as my assistant, we published the *International Journal for the Study of Animal Problems*. This quarterly journal evolved into *Advances in Animal Welfare Science*, an annual review published from 1984-1987. These publications did much to foster a scientific approach to determining animal suffering and well-being, to establishing humane standards of care, and to analyzing attitudes toward animals and the human-animal bond. They also encouraged the scientific community to acknowledge the significance and practical relevance of the science of animal welfare and helped break down institutional resistance to a subject that was hitherto based on primarily subjective criteria, emotionalism, and abstract philosophical rhetoric.

Several veterinary colleges and other universities with animal science and related degree programs quickly incorporated animal welfare science and ethics into the teaching curricula. As a veterinarian and ethologist (a scientist who studies animal behavior), I recognize the limitations of the scientific method, based on deductive reasoning, especially when it comes to evaluating animal suffering and well-being. The clinical method, based on inductive reasoning, and also on intuition, empathy, and common sense, is also important. So too is the application of basic bioethical principles to our regard for and treatment of other sentient beings. The scientific method alone cannot give us all the answers. It concerns me deeply that most government policies and regulations are purportedly "science based" concerning how animals and the natural environment are treated, and new technologies such as genetic engineering biotechnology are approved. But the kind of science used, what I call "scientism," is extremely questionable, not only because it is linked with economic and other subjective values and vested interests, but also because it has become sanctified as truth. Policies and regulations cannot be adequately formulated on the basis of such scientific rationalism in the absence of ethics and common sense, otherwise the consequences could be harmful and absurd. Some of the "science-based" regulations and standards currently on the books for animals in laboratories, for the transportation of farm animals, and for the commercial "puppy mill" production of purebred dogs, for example, clearly reveal the limitations of scientific rationalism.

Q5. DEFINING SUCCESS

The ultimate goal is the transformation/evolution of the human species to a more fully human, empathic, and ethical life form with the boundless ethic of compassion and the principle of equalitarianism being the basis for humane, sustainable, and socially just communities, institutions, economies, bioregions, and nation-states.

Q6. GROUPS & ORGANIZATIONS

I network with many nongovernmental organizations concerned with public health and consumer rights, environmental protection and health, and animal protection and liberation.

Q7. "THE OPPOSITION"

I regard as the opposition all people and groups who place people before animals and nature, and who in policy and praxis manifest anthropocentrism, and who see animals and nature as commodities/resources created for man's use.

John Kistler, the editor of this book, asked me to write about my adversaries, those whom I see as standing in the way of real progress in animal protection (ideally, liberation) and wildlife conservation. My adversaries to such progress are those values and perceptions that have an economic, ideological, and attitudinal basis that is utilitarian and anthropocentric. This rational materialism of custom and commerce sanctifies exploitation of animals. Conscience is assuaged by the setting up of welfare codes and regulations as for puppy mills, livestock transportation, and for considering the possibility of enlarging cages for laying hens or for dogs and monkeys in government and corporate biomedical and military laboratories.

All decisions, in terms of animal use, be it how much pain an experimental animal should be exposed to, how best to medicate to reduce pain, or how many whales can be killed and "harvested" sustainably, are "science based." Thus, the worldview of rational materialism uses science to sanctify and preserve the status quo for those values and virtues that are the enemy of animals and of our humanity. Whales continue to be harpooned, elephants beaten and chained, as billions of hens, pigs, and other innocent creatures are drawn into the holocaust of global industrialism and consumerism.

The enemies of the animals are mine: They are the vices of arrogant anthropocentrism and ignorant self-interest. Along with the values and implicit assumptions of rational materialism, these vices create a spiritual chasm between those who feel a close kinship with all life and those who contend (like the Stoics) that animals are inferior, irrational, have no rights, no interests, or feelings or souls, or higher purpose other than to serve humankind.

These vices and values are like malignant cancer cells to the human soul. Now, through the World Trade Organization (to its infinite peril), they are rapidly metastasizing, as once humane and sustainable plant, animal, and human communities are being supplanted by the monoculture of an industrial economy that justifies global rape and the pillage of the last of Nature's resources, biodiversity, and cultural diversity in the name of progress. Indigenous peoples and sustainable economies, their cared-for animals and fields are being displaced, and, along with wildlife and wild lands, are becoming extinct.

There are many who care and who could really do something to help save the last of the tribals, the elephants, and other endangered species, and all that is sacred to so many—their families and hills and forests—and stop the industrial holocaust of billions of domestic animals. But they do not really do anything beyond regulating and occasionally prosecuting animal exploiters and environmental desecrators, or conduct more scientific research. Through philanthropy (i.e., anthrophilia), rather than biophilia (love for all life), the charade goes on, as human and nonhuman suffering and extinction rates of indigenous cultures and wild plant and animal communities intensify.

It is not so simple as governmental and nongovernmental groups (i.e., animal protection and conservation organizations) being corrupted by money and vested interests, though that is a problem. Rather it is the endemic, cross-cultural chauvinism toward other sentient beings that is our own soul-corrupting element. This creates the chasm between those who love and care for other sentient beings, and those who are able to rationalize how not to care and be responsible for all their "science-based" policies, actions, and their terrible consequences.

I am especially outraged, justifiably as a scientist and veterinarian, over the complicity of animal production and management scientists in the development and continuation of "factory" livestock farms. I mourn that the animal husbandry professional and livestock veterinarians have become animal production and management scientists, biotechnologists indeed, in the wholesale exploitation of animals for human consumption and at such cost and loss in terms of human health, wildlife, biodiversity, and rural life, and once sustainable local economies.

Such scientists are goal directed, and their funded task is to improve the productivity and efficiencies of animals caught in the industrial holocaust of the food and drug industry. It is not for them to change or challenge the system. Their science—scientism or scientific determinism—is linked with the economic and cultural determinism of the ruling industrial-political elite that determines tenure, the freedom of the press, and what children shall be taught.

Scientism, like economism and religionism, is a system of rationalization and denial that discounts externalities or hidden costs—like the cost of such a cruel existence in the service of humankind of billions of animals exploited worldwide on the "factory" farms.

Scientism in all its full colors—rather like the emperor's new clothes—was on display at a conference I recently attended on the plight of the Asian elephant. I asked one speaker, an Indian scientist who is a world-renowned expert on Asian elephants, if he would be the catalyst to help put an end to the traditional culturally accepted cruelties of elephant management in India. He could do this by encouraging adoption of some of the humane alternatives that have been developed in the West as an alternative to the starving, beating, and constant chaining of these animals. I offered to play an audiotape of an elephant screaming while two men were beating him for forty-five minutes on his wounds. (For further details, see *www.gcci.org* and click on "Loki.") Instead of the scientist saying that he cared about elephants and would do whatever he could, he said he would "need more scientific documentation." What more evidence does a rational and compassionate human being need to know than that the screams of elephants being beaten means suffering?

To conclude, people are of two minds. There are two kinds of people, those who would rather give than take, and those who take from life regardless. How these two worldviews work things out in the future is a matter of conjecture, because by then I will be dead. Suffice it to say that a healthy mind is a balanced mind–and world. We of two minds, unlike other animals, must heal ourselves, become of one mind again, save the animals from the holocaust, and also our humanity, all that makes us human.

Q8. HEROES

I get a lot of inspiration from organic farmers, from my jungle tribal friends in India, and from the writings of Lao Tzu, Tolstoi, Teilhard de Chardin, Black Elk, Thomas Berry, and Wendell Berry, to name a few.

Q9. RELIGION/SPIRITUALITY

Spirituality is the center of my life. Childhood experiences provided the basis for what I subsequently understood to be the realization of panentheism—that divinity is in all, and all is in divinity. Panentheism is the spirituality that sees God or divinity in all, and all in God, as distinct from pantheism and animism that sees God or spirit in a particular tree or other living entity. I have discussed this spirituality in detail in my book *The Boundless Circle*. This spirituality is a source of motivation, inspiration, renewal, and self-affirmation (especially when I am with animals or in nature) which helps me continue to help people realize what it means to be human and to realize that all of life is sacred.

Q10. FUNNY/STRANGE EXPERIENCES

I think the following experiences are funny in a macabre sense because they reveal the Orwellian "double think," "newspeak" rhetoric of the dominant culture.

1. Animal scientist Dr. Stanley Curtis told a Congressional Subcommittee hearing on veal calves that "there is no scientific evidence that veal calves need to turn around."
2. Veterinarian Dr. Thomas Wagner, who was engaged in the genetic engineering of cows, told *Fortune* magazine that "a cow is nothing but cells on the hoof." (October 1987, p. 80).
3. After a Senate Subcommittee hearing on the cloning of Dolly the sheep and possible human cloning, a reporter for *Science* magazine said to me, "It is disturbing how far people will go to maintain their depraved existence."

My most rewarding experience: Having a farmer come up to me at a conference and say he heard me speak ten years earlier about farm animal welfare and factory farms and at that time he was operating a hog confinement system. What I said made him change, and he became an organic farmer and "never looked back."

My most moving experience: Sometimes when I am treating an animal, I feel in my own body where the animal is suffering, which leads me to appreciate the power of empathy and its diagnostic and healing potential.

Q11. LESSONS/ADVICE

Read, learn, be informed, be prepared to suffer but find the strength of your convictions and your own source of healing and renewal in the communion of kindred spirits, including those who are not in human form. Try to avoid being judgmental and show compassion and respect toward those who are inhumane, ignorant, and indifferent. Acknowledge the fact that society is dysfunctional and that spiritual corruption is the major disease of our species that no amount of animal experimentation is going to cure. Until we are in our right minds, heart-centered, we (and the world) will never be well. I have learned, and am still learning, to deal with my rage and disappointment. I am learning not to react in anger, or to be judgmental or hateful of those who do not feel as I do: that we should give all sentient beings equal consideration and that human liberation means animal liberation and the protection and restoration of the natural world.

Various organizations and individuals in the animal welfare/protection/rights-liberation movements have different strategies, policies, principles, and perceptions. There can be strength in diversity, provided we avoid in-fighting, ideological rigidity, and irrational fundamentalism/fanaticism. A common ground and a common cause under the umbrella of reverential respect for all life can unify the movement. However, the future and strength of this movement will only come, I believe, when there is a stronger bioethical, spiritual, and political linkage between human well-being, Earth care, and respect for the intrinsic value of all sentient beings.

Q12. ADDITIONAL INFORMATION

A life unexamined, as Socrates said, is a life unlived. If we have nothing worth dying for, then what are we living for? We need to redefine/rediscover what it means to be human and to realize our own inner divinity and the power of love, rather than the love of power that is so corrupting of our humanity and of the life and beauty and the mystery of Creation.

Milton M.R. Freeman

Q1. BIOGRAPHICAL PROFILE

I live in a medium-size city on the Canadian prairies where I taught anthropology until my retirement in 1999. I have three adult children and five grandchildren.

I was born in London, England, in 1934, but my family moved to the countryside of southern England when I was five years old. As a boy, I made collections of animal skulls, birds' eggs, insects, and rocks, and kept a number of live animals—including fish, tree frogs, newts, snakes, lizards, and songbirds. Although I had an air rifle, I only shot at nonliving targets. I remember, as a child, going on an antimeat campaign (for a few days) after I found one of our backyard chickens dead on the kitchen table. I had thought we only kept chickens for their eggs!

As a young high school student in England, I enjoyed reading about the Arctic. I was attracted by the "romance" of that distant region: accounts of polar explorers, fur traders, and local peoples, not to mention fascinating and exotic animals such as polar bears, musk oxen, whales, and walrus. These northern interests influenced me to study science in high school and university, for it seemed to me that becoming a scientist was a good way to get to the Arctic. While still an undergraduate student, I seriously considered spending a season as an inspector on a British Antarctic whale ship, the *Balaena*. However, being prone to seasickness, I abandoned that particular plan.

However, now that travel involves less ship time than it once did, I enjoy the different cultural experiences that travel offers—whether in Canada or in other countries.

Q2. BECOMING INVOLVED

Working with Inuit in the Arctic, and living for some years in Newfoundland at a time when the seal hunt was also a vibrant part of that rich regional culture, I was angered by the victimization of sealers by radical environmentalists and animal rights activists taking place in the 1970s. But what incited me to actively oppose animal rights activities occurred a few years later, when I started attending the annual meetings of the International Whaling Commission (IWC).

One big issue at the IWC in the early 1980s was to battle to "save" the Bowhead whale. Biologists, at that time, believed that Bowhead whales in Alaska were at serious risk of being driven to extinction by the Inupiat whalers' traditional hunts. However, Alaskan whalers believed that their hunting posed no threat to whale stocks. The IWC, having by then transformed itself into an antiwhaling organization, listened only to the biologists. The result was that those possessing only wrong information misled those who knew nothing of the lives and knowledge of the Alaskan whalers (but who wanted to end whale hunting) with an inevitable, but unjustified, decision to stop the Bowhead hunt.

Naturally the whalers were shocked and refused to accept this decision. Thus I became introduced to the politics of "whale saving" and exposed to the fervor of animal rights (and radical environmental) activists jumping on the "Save the Whale" bandwagon which was beginning to gain momentum.

Attending annual meetings of the IWC provided the opportunity to observe the skill and fervor with which animal rights activists played on the emotions of delegates, the media, and the politicians. As occurs often when an animal rights campaign chooses to target a people for attack, or a photogenic animal species for protection, the truth is often the first casualty.

Over the past twenty years, I have seen animal rights activists (and the journalists and decision makers in government who uncritically accept what activists tell them) act as though they know what is best for people they know so little about. It is this sustained attack on others' human dignity and total well-being that fuels the activism in others, including myself, who believe that people have rights, that these rights are being trampled, and that human lives and worth are being trashed.

Q3. IMPORTANT ISSUES

Over the past thirty years, I have been involved in trying to educate people about the social, economic, cultural, and health-related circumstances of diverse coastal peoples whose livelihood and food comes from marine mammals. The reason I have focused my activities upon these issues is because for most people living in urban environments whose food comes from supermarkets, the hunting of whales and seals appears anachronistic and needlessly barbaric.

But rural people have fewer choices for making a living than do city dwellers. These coastal people, who often live in relatively remote small communities, feel bound to their occupation and dwelling place by very profound and meaningful bonds of history, family, tradition, and love of home place. Furthermore, they see a virtue in providing wholesome food to their families and to others—much as farmers value their chosen occupation and ties to the land.

Working in the Arctic and in Japanese coastal villages, and visiting other whaling locations in Europe and Oceania, I have witnessed the profound damage and despair caused to families and communities by animal rights activists. I understand that city dwellers who choose not to eat meat or take no active part in obtaining their own food from nature may find it very unpleasant to contemplate killing an animal in order to feed themselves and their families.

However, this squeamishness gives such people no legal or moral right to devastate the lives of hunting peoples who live in quite different, and often much less-privileged, circumstances. The men, women, and children of these rural societies have built their sense of worth as human beings, indeed their very identity over many centuries, on particular ways of life and cultural beliefs that provide them with a fulfilling and moral life. As agriculture often cannot be practiced in these northern or rocky regions, human life is necessarily sustained by hunting and fishing (and sometimes trapping) wild animals.

For some people, raising their children close to nature and away from city dangers is a conscious choice, and one that every human being has the right and responsibility to choose for their own children. Yet, if denied an opportunity to earn a living from the local resources, then the right to exercise that choice is denied. Simply put, actions of others to take away this choice is denial of a human right, and it is precisely this denial of human rights that results from the actions of animal rightists who oppose the hunting of whales and seals.

Do I consider myself an activist? Let me say that there are some practices involving the treatment of human beings, in my own society as well as elsewhere, that I feel are morally wrong, and within the legal and socially accepted rules of dissent I actively support those working to redress these problems. This is especially the case if I believe injustice or unwarranted harm is being caused—or likely to be caused—to innocent people. Thus, I oppose capital punishment, child prostitution, and other forms of imposed servitude, racial and sexual discrimination, and animal rights. In that sense I consider myself an activist.

Q4. GOALS & STRATEGIES

Short-term, I work with like-minded people in defending the rights of people needing to use nonendangered species of animals in a sustainable manner. As these resources are primarily used to feed people and to support the social well-being of their communities, I consider the work I am engaged in to be one of continuing urgency and importance.

On a long-term basis, I hope to assist in providing educational material to counter what I consider to be very biased information presently made available by those who place animal welfare ahead of human welfare. This I see as a dangerous trend that needs to be strenuously countered.

Although now retired, I continue writing and speaking about the environmental and social benefits associated with the sustainable use of natural resources. For many years I have organized and participated in international conferences and workshops that focus upon resource management issues, or the sustainable use practices of marine hunters. I continue with these activities. I also occasionally write "letters to the editor" of mass circulation newspapers, usually in response to what I see as erroneous or biased reporting, especially in relation to whaling, which I consider to be badly misrepresented to the public.

Q5. DEFINING SUCCESS

The ultimate goal is to have an informed public, one that can look critically at contentious issues and know when emotion is clouding rational thinking and when artifice is being used to obscure or distort the facts.

When would I consider the task complete? I would answer this by asking: When will the work of schools and colleges be finished? Obviously the answer is never: the need for education is ongoing. Today, animal rights is part of a multimillion-dollar protest industry that has established its occupational niche in our affluent, information-based, urbanized world. Just as the tobacco industry, drugs, or violence in the Hollywood film industry will continue to exist and cause harm, so will the animal rights movement similarly strive to prosper at others' expense. As a result, those who see the need to oppose the harm caused by such activities will also have every reason to continue working to counter the damage.

Q6. GROUPS & ORGANIZATIONS

I work with such organizations as the Inuit Circumpolar Conference (representing the interests of Inuit sealers and whalers and their families in Canada, Greenland, Russia and the United States) and the World Council of Whalers (representing whaling peoples across the Arctic, in the North Atlantic, the Caribbean, Southeast Asia, and the North and South Pacific regions). I also participate in sustainable resource use activities of the International Union for the Conservation of Nature (IUCN) and the International Wildlife Management Consortium (IWMC), which are two global coalitions involving governments and nongovernmental agencies that promote the sustainable use of natural resources through research and education.

Q7. "THE OPPOSITION"

I fully support efforts to increase animal welfare, strongly believing that no animal should be treated cruelly or made to suffer unnecessarily. But I do not believe it is wrong to kill an animal for food, or to use its skin or other parts for commercial or other purposes, if the amount of distress caused the animal is minimized.

I may even agree with animal rights activists on one issue, believing a wild animal in nature is more likely to have a better "quality of life" than most farm animals do. However, unlike animal rightists, I believe this is so even if the wild animal spends the last few minutes of its life being killed for dinner. In this regard, I believe animals often fare better at the hands of human hunters than they do at the teeth of a natural predator which may start eating its prey while it is still alive. Similarly, I do not believe all animals dying a "natural death" enjoy an easy death. I ask myself: How many hours or days does a dying whale struggle to stay afloat before it drowns? Indeed, I suspect a skilled hunter's bullet may provide the most humane way for an animal to end its life.

Q8. HEROES

In high school I became attracted to the activities of leading pacifists, some of whom, like Bertrand Russell, had been jailed for their beliefs. Others I admired for their nonviolence and selflessness were Albert Schweitzer and Gandhi. Clearly, Martin Luther King Jr. is a recent exemplar of those virtues. Being attracted to the exploits of polar explorers, I greatly admired Fridtjof Nansen, for he never lost a man on any of his hazardous polar expeditions. In addition, I greatly admired this great Norwegian's humanitarian actions in saving the lives of tens of thousands of Armenians fleeing slaughter as world governments did nothing.

Q9. RELIGION/SPIRITUALITY

Although not a religious person myself, I recognize the wisdom in many sacred teachings, and also the need for moral order to be followed in society. As an omnivore, I do not personally subscribe to the Hindu belief against eating beef, nor the Jewish or Muslim belief that pork is unclean. But I respect those possessing these religious beliefs, and I consider this respect and forbearance to be essential if we are to live together, with harmony, in a multicultural world.

I have a tendency (shared with many other people) to oppose those who set out to deliberately harm others. I find that this is my motivating or sustaining force in opposing the damage animal rights beliefs inflict on people. As animal rights activists often ignore or otherwise devalue the spiritual beliefs of, for example, Buddhist whalers in Japan or Inuit whalers in the Arctic, I feel especially motivated to speak out against what I consider to be indefensible bigotry and religious and spiritual intolerance.

Q10. FUNNY/STRANGE EXPERIENCES

I have decided to devote the available space to matters I consider more important to comment on. This is not to say that there were no funny or strange experiences—merely that I am conscious of having used up my allotted space.

Q11. LESSONS/ADVICE

I have learned that animal rights activists are very dedicated to their mission and that they are skilled campaigners. On the other hand, I believe that most people do not want to hurt other people, and most possess a sense of fairness. The inherent weakness of the animal rights position is that it sets out to cause harm to people who themselves cause harm to no one. In fact, the victimized rural people possess a number of virtues that most people admire: they are hardworking, family centered, self-sufficient people whose lifestyle causes minimal environmental damage.

Twenty years ago I made the serious mistake of underestimating the degree to which the animal rights activists' emotional campaign would win the hearts and minds of the public, and particularly the media. It is now an uphill battle to gain media attention with rational arguments, for example, about the conservation benefits of sustainable resource use. The answer might be to fight fire with fire, to appeal to emotional issues.

For just as people's hearts melt at the sight of a wide-eyed seal pup or a whale calf with its mother, so people can feel love for a wide-eyed innocent child, or empathize with any mother's tender affection for her children. All people have indisputable rights—and innocent victims elicit widespread empathy. That, I believe is the Achilles' heel of the animal rights movement, for with an informed public, the needs of children will always trump animal rights for the great majority of rational people.

Q12. ADDITIONAL INFORMATION

As an anthropologist and a conservationist, I have tried to share my understanding of the human-environmental relationships that continue to be important to various peoples' health and well-being. Although concerned about environmental problems that many others worry about, I do not see the solution to these important local and global problems coming out of an animal rights philosophy or action plan.

On the other hand, being interested in conservation (meaning the rational use) of natural resources, I accept that for animal populations to remain healthy and flourish, management practices may require that some individual animals may have to be removed so the population of animals as a whole remains healthy. Just as many believe that human overpopulation threatens our collective future survival, so wildlife managers and conservationists believe there are optimum population levels for animals too, and that the food and habitat required to keep wild animals healthy can eventually be used up by an ever-expanding population of wildlife.

Margery Glickman

Q1. BIOGRAPHICAL PROFILE

I was born in Brooklyn, New York, in 1947. In 1968, I graduated Phi Beta Kappa from Hunter College with a major in Philosophy and a minor in Political Science. I have an M.E. degree in Curriculum and Instruction from Loyola University, and I taught in public elementary schools both in Illinois and Florida.

In 1977, I married Fred Glickman. We have three children, David, Michael, and Laura. After being diagnosed and treated for breast cancer in 1990, I advocated for increased funding for breast cancer research.

I enjoy reading, the company of dogs, watching movies, walking, swimming, advocating for animal protection, and watching wild birds.

Q2. BECOMING INVOLVED

My interest and involvement in animal rights issues grew gradually over the last twenty-five years. My efforts became focused on helping the Iditarod sled dogs after a 1998 trip to Alaska during which I visited Iditarod dog kennels, including ones owned by Iditarod race winners. I was horrified to see dogs permanently tethered on four- or five-foot chains as their primary means of enclosure, living in their own fecal material, and drinking water from rusty cans.

Q3. IMPORTANT ISSUES

Most of my time is spent working against the Iditarod. This is a brutal dog sled race in which dogs are forced to run 1,150 miles over grueling terrain in a time frame of nine to fourteen days, which is approximately the distance between Chicago and Houston, Denver and Los Angeles, or New York City and Orlando. Dog deaths and injuries are common in the race. Sports columnist Jon Saraceno wrote in *USA Today* on March 3, 1999, that the Iditarod should be called "Ihurtadog," a "travesty of grueling proportions."

The Iditarod dog sled race violates accepted standards regarding animal cruelty as shown by the laws of thirty-eight states and the District of Columbia. Those laws say "overdriving" and "overworking" an animal constitutes animal cruelty. The California law is typical:

> 597. Cruelty to animals.
> (B) Every person who overdrives, overloads, drives when overloaded, overworks . . . any animal . . . is, for every such offense, guilty of a crime punishable as a misdemeanor or as a felony or alternatively punishable as a misdemeanor or a felony and by a fine of not more than twenty thousand dollars ($20,000). (Animal Welfare Institute)

The dog deaths and injuries in the Iditarod show that these dogs are "overworked" and "overdriven." If the Iditarod occurred in any of these thirty-eight states or the District of Columbia, it would be illegal under the animal cruelty laws. Unfortunately, the Alaska's animal anti-cruelty law does not say that "overdriving" and "overworking" an animal is animal cruelty.

Dogs have died in the Iditarod from strangulation in towlines, internal hemorrhaging, liver injury, heart failure, pneumonia, "sudden death," and "exertional myopathy," a condition in which a dog's muscles and organs deteriorate during extreme or prolonged exercise.

According to Tom Classen, retired Air Force colonel and Alaskan resident for over forty years, the dogs are beaten into submission. "They've had the hell beaten out of them." You don't just whisper into their ears, 'OK, stand there until I tell you to run like the devil.' They understand one thing: a beating. These dogs are beaten into submission the same way elephants are trained for a circus. The mushers will deny it. And you know what? They are all lying" (*USA Today*, March 3, 2000).

The Iditarod spawns puppy mills. In Iditarod kennels or puppy mills, killing unwanted dogs is a common practice among mushers. These mushers breed many dogs, hoping to get a few that will be fast enough to race. According to an article in the *Anchorage Daily News*, "Killing unwanted sled-dog puppies is part of doing business" (October 6, 1991). Most of the mushers cull by shooting their dogs in the head. An animal that is not properly restrained when the musher fires the gun may suffer an agonizing death.

Mushers also kill dogs that are injured in the Iditarod: old, but otherwise healthy dogs; or any dog that is no longer wanted for any reason. According to Mike Cranford of Two Rivers, Alaska, "[O]n-going cruelty is the law of many dog lots. Dogs are clubbed with baseball bats and if they don't pull, dragged to death in harness." (Quote taken from *The Bush Blade Newspaper*, serving Cook Inlet and Bush, Alaska, March, 2000, in a Web site article.) (Imagine being dragged by your neckline at fifteen miles per hour.) Musher Lorraine Temple said, "They (the big racing outfits) can't keep a dog who's a mile an hour too slow" (*Currents*, Fall 1999).

Dogs return from the race to kennels where they are kept permanently tethered on chains that can be as short as four feet long. Tethering is cruel and inhumane because:

Continuous chaining psychologically damages dogs and makes many of them aggressive animals.

A dog that is permanently tethered is forced to urinate and defecate where he sleeps, which conflicts with the dog's natural instinct to eliminate away from his living area.

Because the chained dog is always close to his own fecal material, he can easily catch deadly parasitical diseases by stepping in or sniffing his own waste.

Even if the fecal matter is picked up, the area where the dog can move about becomes hard-packed dirt that carries the stench of animal waste. The odor and the waste attract flies that bite the dog's ears, often causing serious bloody wounds and permanent tissue damage.

In 1997 the U.S. Department of Agriculture determined that the tethering of dogs was inhumane and not in the animals' best interests. The chaining of dogs as a primary means of enclosure is prohibited in all cases where federal law applies.

Q4. GOALS & STRATEGIES

My short- and long-term strategies are the same. I educate, protest, write, and research on a regular basis, but most of my efforts take place in the several months before the start time of the Iditarod. I developed the Web site *www.helpsleddogs.org*, as a focus of communication and education.

Everyone should be aware that mushers are using the public schools (and thus our tax dollars) to transmit their message of animal cruelty to unwitting schoolchildren, who are unable to see behind the veneer of sports heroism and ask the important questions about animal welfare associated with this public spectacle. In fact, schoolchildren are a primary target of marketing by the Iditarod Trail Committee, which generates teaching materials that have everything to do with promoting propaganda, but ignore or minimize any discussions of the race's toll on the participating dogs. Every parent and taxpayer should be aware of this situation and concerned that it violates many states' requirements for humane education.

Q5. DEFINING SUCCESS

I believe that success should be looked at incrementally. Every stride that results in improving the lives of the dogs or in educating people about the brutalities of the race and the kennels is important.

Thanks to the efforts of many people working on behalf of the dogs, many companies have disassociated themselves from the Iditarod. These include Avon, Novartis, Hill Brothers Brand (Nestle), Rite Aid, Outback Steakhouse, Carr-Gottstein Foods (Safeway), Maxwell House Brand (Kraft Foods), and Parker Brothers. A complete list of our successes can be found on *www.helpsleddogs.org/success.htm*, our Web site.

Q6. GROUPS & ORGANIZATIONS

I am the founder of the Sled Dog Action Coalition. The coalition is composed of organizations and individuals dedicated to helping the dogs forced to run in the Iditarod dog sled race and the dogs forced to live in the Iditarod kennels. We are opposed to competitive mushing events that sacrifice the welfare of participating dogs for the sake of human glory, enhanced competition, and entertainment. The Sled Dog Action Coalition does not solicit or accept money.

The Iditarod dog sled race is condemned by animal protection groups across the United States. As director of the Sled Dog Action Coalition, I work with a large number of organizations. These include the Humane Society of the United States, People for the Ethical Treatment of Animals, Association of Veterinarians for Animal Rights, In Defense of Animals, United Animal Nations, Last Chance for Animals, Animal Protection Institute, Ark Trust, the ASPCA (American Society for the Prevention of Cruelty to Animals), and the Doris Day Animal League.

Q7. "THE OPPOSITION"

In general, I consider the opposition to be anyone who supports the Iditarod. More specifically, I include my opponents to be:

Iditarod Trail Committee members

People who have raced or who want to race their dogs in the Iditarod

Teachers who promote the race to their students

Veterinarians who serve as Iditarod publicists by handing out pro-race propaganda

Husky dog owners who support the race

People in the media who are propagandists for the race

Companies that endorse the race

Q8. HEROES

I was, and still am, inspired by my late beloved golden retriever, Cornflake, who had limitless love and goodness.

Mikayla "Kayla" Schramm, a thirteen-year-old honor student and ardent animal protection activist, also inspires me. From the time she was a baby, ignoring fire ants in order to save some earthworms, Mikayla has shown steadfast compassion and dedication in helping animals. As a small girl, her nose was broken by neighborhood bullies who resented her for saving little fish and crabs from a drainage ditch.

Mikayla was featured on the front page of the *Texas City Sun* concerning her correspondence advocating animal protection. Months later, a follow-up article acknowledged her writing to a district attorney in Kansas City, Kansas, about the trial of four young men who tortured a small dog named "Scruffy" to death. Her letter was read in court where it reportedly helped the judge choose the highest punishment possible for each of the young men. The district attorney later called Mikayla to thank her for her unsolicited assistance, and to advise her of the trial's outcome.

Some of Mikayla's other accomplishments are summarized below:

- Mikayla was acknowledged by an article in the *Tulsa* (Oklahoma) *World* newspaper's front page for writing against ostrich races.
- After having won a national essay contest, "Why Animals Don't Belong in Circuses," sponsored by United Animal Nations, Mikayla was pictured on the front page of the *United Animal Nations Newsletter* and on its Web site.
- Mikayla's letter writing efforts to help animals was published in PETA's Children's Newsletter, "GRRR!" and acknowledged on the PETA Web

site. Mikayla's letter to the editor of *TEEN* magazine opposing the Iditarod was published. She received responses to her letters about animal cruelty from Vice President Gore, congresspersons, senators, mayors, county commissioners, district attorneys, and prime ministers.

- A Web page with her comments concerning the Iditarod was posted on the "Help Sled Dogs Action Coalition" Web site, *www.helpsleddogs.org*.
- At age eleven, while alone in the woods, Mikayla recaptured an adult male Canadian Timber Wolf who escaped from a sanctuary. It was the start of hunting season in Texas, and without her brave actions, the wolf would likely have been shot, starved, poisoned, or hit by a car. Mikayla was adopted by the Cherokee Nation and given the name "Little She Wolf" in an official naming ceremony for her feisty spirit in saving the wolf.

Q9. RELIGION/SPIRITUALITY

Spirituality is intricately woven into the fabric of my life and this helps motivate my work. I have the utmost respect for life.

Q10. FUNNY/STRANGE EXPERIENCES

[No response.]

Q11. LESSONS/ADVICE

Carefully and intensively research the topic, so you have all the facts. Provide quotes that prove your points.

Do not be deterred if your opponents respond with vicious attacks and make threats against you.

Thank the activists who have helped you in your campaign. If you have news of a good result, make sure to tell them.

Ask activists to send protest emails to the respective company investor relations e-mail address.

Create a Web site and include pictures and quotes when appropriate.

Join e-mail lists that support the opposition's point of view.

Q12. ADDITIONAL INFORMATION

As I work to help the Iditarod dogs, I am often reminded of a scene from *All the President's Men*, the famous book and movie about Watergate. The anonymous informer Deep Throat tells *Washington Post* reporter Bob Woodward to "follow the money," when he asks for further clarification for his investigation into the Nixon White House. This phrase makes it easier for me to understand the entire Iditarod culture and how so many people can look the other way to justify an event that is harmful to so many dogs that the mushing community claim to love and responsibly care for. The mushers, their sponsors, the Alaskan media, Alaskan politicians, and Alaskan businesses all benefit financially from this dog-killing spectacle. I urge other animal activists to keep the money trail in mind while conducting careful research into the activities that they seek to stop.

Kimber Gorall

Q1. BIOGRAPHICAL INFORMATION

My name is Kimber Gorall. I was born in February 1957, the sixth of nine children. I grew up in a rural suburb in northwestern New York State. My elementary school bordered a dairy farm, and I spent recess each day petting the cows and feeding them grass. My family lived in a very small house with a large backyard. I remember swinging on vines and climbing tall willows, building tree forts, catching tadpoles, and skating on a makeshift pond.

I would classify us as working poor. My father was a brilliant though very underpaid self-taught chemist. My mother stayed home with the children. She was in poor health for much of my life, and she died when I was in college, leaving three children still at home. Ninety-nine percent of who I am I attribute to my father. But the other 1 percent, and perhaps it's the most important, is the need to stand up for what I believe in, no matter what risks I face in doing so. Undeniably, that trait comes from my mother.

As a child, people (of all ages), languages and cultures, animals, plants, rocks, and galaxies fascinated me. By about sixth grade, I stopped worrying about being popular, and focused instead on being myself. When I was in seventh grade, I created an environmental activism group at my school. That was the year of the first Earth Day.

After high school, I earned a degree in Geological Sciences and a Certificate in Telecommunications Management. I have worked in a wide variety of fields, in both blue- and white-collar jobs. I was a mineralogist in Bisbee, Arizona. I had my own subchapter S corporation for five years. Currently, I am manager of corporate communications for an Inc. 500 software company.

Kimber Gorall. Photo credit: Lisa Emerald Kaufman

People who know me say that I am passionate, driven, and an independent thinker. I view myself as both an idealist and a realist. I am no doubt a dreamer, but I am also practical, and I believe I have both feet on the ground. I enjoy being productive at my job. I also enjoy writing, music, ice skating, outdoor activities, cooking, and baking and spending time with friends. Activism, though, is a priority in my extraprofessional life.

Q2. BECOMING INVOLVED

It was both a single event and a long process that brought me to activism!

Twelve years ago, I was living alone in a big house, and after a frightening experience, I decided to get a dog for protection. I wanted a "quality" dog, so I went to a pet store to look at purebred puppies.

There was one dog who looked truly depressed, and a burly man was teasing him. I told the man to leave my dog alone, and I took the puppy home. I had recently given away my television, so when I got home from work each day I had nothing to do except play with "Spike."

A few things quickly became apparent to me. I was the center of my dog's universe. Given that his life expectancy was perhaps a dozen years, a single day of my life was equal to an entire week of his. I felt an enormous responsibility to ensure his quality of life. Most importantly, though, I began to realize how intelligent, sensitive, and perceptive my dog was. I found myself reexamining everything I'd ever been taught about the way animals should be treated. Inevitably, I questioned why we treat some animals with reverence and others with no respect whatsoever.

Like so many other people, I had always professed to care about animals. Throughout my life, whenever I had seen cruelty toward animals, I had tried to stop it and spoke out vehemently against it. I financially supported the local zoo because I thought it was doing good and necessary things for animals. I went to the aquarium, fairs, and the circus because I enjoyed the company of animals. Yet I dissected worms and grasshoppers in Biology class. I wore fur, leather, wool, and silk, which I considered to be natural materials because they were derived from nature. I ate any and all animal foods. Although I would never consciously harm an animal, I realized that I was paying others to do it. I was a hypocrite.

When I decided that I could no longer justify eating meat, I knew that I must ultimately become a vegan in order to be true to my beliefs. Health was never a concern for me. I knew only that I didn't want animals to be killed to satisfy my appetite for their flesh. Having grown up on the standard american diet, it took me a while to figure out that I could in fact live more healthfully by not consuming any animal products whatsoever.

Eventually, I felt compelled to do something outside my little world, to channel my anger and grief into action. I remember the first thing I did publicly as an activist. I joined two hundred others at a walk in a public park, organized by opponents of a controversial program in which deer were baited and shot. A walk in the park—I was comfortable with that.

The next day the front page of the local newspaper featured a large color photo of me and my dog. The caption included our names and the word "protesters." I was mortified. Now I no longer cringe at labels.

My dog Spike is still with me, though he is old now. What a beautiful little spirit he is! He is the embodiment of love and joy and charity and wisdom and tolerance—of all things good. He is the inspiration for all that I have done and will do as an activist. He forever changed my life in a way that no human could have done. He awakened my consciousness by giving the plight of animals in our society a face, a personality, a life story.

Q3. IMPORTANT ISSUES

Two issues that I care deeply about are vivisection and factory farming. I think they trouble me more than most because animal experimentation and animal foods are promoted in our society as necessary to ensure our health. Yet there is a reason why people aren't allowed to see behind the closed doors of laboratories and factory farms and slaughterhouses. They are among the most hideous of the animal abuse industries. Millions of animals are tortured in laboratories each year. In this country alone, 9 billion animals are slaughtered each year for their flesh.

If people could witness the miserable lives and cruel deaths these animals suffer, public opinion would probably force such places out of business.

I am a vegan evangelist. I spend a lot of my time promoting a plant-based diet. I believe that vegetarianism holds the key to creating a new ethic in our society. It is simply not necessary for us to eat animals in order to live. On the contrary, every day there is more medical evidence reported that animal foods are unhealthy and that a vegan diet is a health-promoting diet. Furthermore, the horrendous conditions in factory farms and slaughterhouses have produced numerous outbreaks of deadly bacteria such as listeria and salmonella, and have resulted in government mandated warning labels on meat. Millions of us have heart disease, cancer, food allergies. The standard american diet is killing us.

Some people fear that giving up meat will compromise their health. But every human baby is a vegetarian during the first several months of its life, that period during which we experience our greatest growth. It's a simple yet powerful step that everyone can take—to stop eating animals.

Q4. GOALS & STRATEGIES

Activism for me is a calling, not a choice. I couldn't avoid it if I wanted to. It seems as though I can't go anywhere without becoming involved in some activity or conversation related to animal rights. Working primarily on a local level, I do anything and everything that I can to help animals.

I have done leafleting, tabling, and protesting at circuses, pet stores, laboratories, slaughterhouses, the rodeo, fur stores, McDonald's and Burger King, Procter and Gamble, hunter check-in stations, aquariums, fairs, zoos and petting zoos, wildlife refuges, government offices, concerts, festivals, health fairs, environmental fairs, humane education events, hunting shows, grocery stores, and schools and universities. I have organized free, public lectures by nationally known speakers. I participate in an annual walkathon to benefit Farm Sanctuary. My dog goes along and has his own pledge card. In 1996, I joined thousands of others in a March for Animals in Washington, D.C.

I helped construct four large, permanent shelter buildings for feral cats. I worked on a monthlong project that entailed using chain saws to cut six-inch steel spikes off the top of a mile-long fence on which deer were accidentally being impaled. A follow-up project involved capping the spikes with flexible hosing along another stretch of fence.

Every Saturday and Sunday for the past four years, I have worked a phone line to the local community. I handle every sort of call, from animal abuse complaints to reports of injured wildlife to people trying to place unwanted pets. I assist students and teachers with school projects related to animal rights. I give interviews, do research for them, and provide literature and other materials pertaining to specific subjects.

Whenever I cook or bake, I try to make extra to give to coworkers and friends. Several have become vegetarian since they met me. My employer allows us to periodically host company breakfasts and potluck lunches. I have always volunteered to bring or prepare vegan food to these events. It is a good way to introduce people to foods they might not otherwise try, and the taste and healthfulness pleasantly surprise them.

Last year, I fasted for more than 150 days on juice and water as part of an international effort to persuade the Food and Drug Administration to remove the recombinant bovine growth hormone, rbGH, from the market.

For the past two years, during deer hunting season, I have donned blaze orange clothing and spent my weekends patrolling a wildlife sanctuary to protect resident deer herds from hunters. The sanctuary is comprised of noncontiguous parcels of land, and they are surrounded by public and private areas that allow hunting. I try to chase the deer away from the hunting camps and into the safe areas, and I report any illegal or threatening activity.

I attended the Fred Coleman Memorial Labor Day Pigeon Shoot in Hegins, Pennsylvania, twice. I was a videographer there. Among other things, I videotaped a man biting the head off a live bird. My footage was to be used in court as an exhibit for the prosecution. I went to court twice. Both times the case was thrown out on technicalities. The organizers of the pigeon shoot then voluntarily cancelled the event for that and future years.

I have secured foster or permanent homes for quite a few cats and dogs who were either homeless or in abusive situations. I currently live with three cats, all of whom are rescues. One of them is feral.

Q5. DEFINING SUCCESS

As someone working at the local grassroots level, I do not have an ultimate goal. I am not a one-issue activist. I am realistic enough to know that I will be in this movement for the rest of my life. We are up against long-held beliefs and prejudices. The industries that abuse animals are well funded. And the problems we are dealing with touch every facet of our society.

Victories will be achieved in small steps. We must change public opinion. Changes in laws and industry practices will follow. This was true for other social justice movements in the past. But unlike previous movements, the objects of our efforts—animals—cannot give speeches or write letters to newspapers, can't vote, can't join our protests.

I don't know if I will live to see the day when animals are granted legal rights. But people once said that about human slaves in this country. I do believe that our society's attitudes toward animals are changing. The issues we care about are getting more attention, and the youngest generation is growing up more aware of them. At protests, we often get thumbs-up from passing motorists, and some stop to engage in meaningful discussion. At tabling events, people approach me and want to know more. I have seen many students choosing animal rights as an essay or project topic.

From personal experience, I have found that compassion can be contagious. One day last summer, for example, I was at the local main post office during a very busy lunch hour. Inside, there was a huge bee throwing itself against a plate glass window, obviously trying to find a way outside. I started looking around for a container large enough to accommodate the insect. Other patrons asked me what I was doing, and before long, there were several people digging through wastebaskets and searching their cars for a container. At one point, I thought, "this is so bizarre. Do these people realize what they're doing?" It was a small incident, but it made me feel good to know that people who normally would not have given the bee a moment's thought got caught up in the spirit of compassion, and tried to help.

Q6. GROUPS & ORGANIZATIONS

I am currently the president of Animal Rights Advocates of Upstate New York, Inc., a 501(3)(c) nonprofit organization. My organization is not formally associated with any other groups, but we do things collectively with other regional or national groups whenever the opportunities arise. I am also affiliated with the Rochester Area Vegetarian Society, Inc.

Q7. "THE OPPOSITION"

It has always been easier for me to see commonalties than differences. I believe that every one of us has the capacity for both good and evil. I also believe that few people are deliberately cruel. So much cruelty is committed thoughtlessly.

Most of us in this movement can say that we were on the other side of the fence at one time. But we changed. Even people who work in animal abuse industries can change. I am reminded of Howard Lyman. To have gone from begin a fourth-generation cattle rancher to an outspoken vegan activist is an astonishing transformation.

To me the opposition is made up of the institutions and the industries that promote and profit from animal cruelty. They rely upon deceiving and manipulating the public to continue doing what they do. I do not condemn the slaughterhouse worker, for example, but I do condemn the slaughterhouse, the factory farm, the meat and dairy industries. They are the real culprits, and both humans and animals suffer for it.

Q8. HEROES

I have always been inspired by people who have beaten the odds—those who have overcome adversity and not only survived, but also thrived. I try to imagine what life must have been like for people animal advocates such as Leonardo da Vinci, Pythagoras, and Francis of Assisi, kindred spirits who were surrounded by ignorance and who lived so far ahead of their time.

In my opinion, this movement is full of heroes. There are pockets of compassion everywhere, ordinary people acting out of consciousness to do whatever they can to make the world a better place. In particular, I have great admiration for wildlife rehabilitators. They are self-taught, unpaid volunteers with no staff and no recognition. They work round the clock to rescue and care for injured or orphaned wild animals. Angels of mercy.

Q9. RELIGION/SPIRITUALITY

I regard myself as a spiritual person. I have always felt a kinship with the whole of creation, and a connection to a wider universe. But it has never been important for me to be affiliated with an organized religion.

My religion, if I have one, is veganism. Ahimsa. When people ask me if I believe in God, I tell them that I am too busy doing God's work to think about such things. I have a very strong sense of purpose, one that is not deterred by what people think of me.

Q10. FUNNY/STRANGE EXPERIENCES

To be honest with you, I can't think of a funny experience. That's not to say that we don't have fun. Most activists I know have great senses of humor. I think we need that, or we'd go crazy.

I do recall one very strange incident, though. It was at the beginning of the deer-hunting season several years ago. I rented some professional animal costumes. I was dressed as a deer, and we also had a turkey and a bear. We rented a four-wheel-drive vehicle. We taped posters with antihunting slogans on the windows, and placed an effigy of a hunter on the roof, complete with beer can in hand, orange vest, and arrows protruding from his back. We drove around the county, getting beeps, giggles, and thumbs-up from passing motorists, and dirty looks from the men outside the hunting supply stores. We knew we'd gotten our message across to the community. It was a very cathartic experience, and we laughed as we rode through town. Suddenly we caught sight of a large buck running into the street right in front of our vehicle. It was a Saturday afternoon, and he ran full speed into heavy traffic, so he must have been running from a hunter. Thankfully, our driver reacted in time, but the deer slammed into the car right next to ours. It smashed the driver's side window and somersaulted over the roof of the car, landing on its feet and running off into the woods. There's no question that the deer was injured, and probably died of its wounds. All of us in the car started crying. Someone asked why this had to happen—were we being punished? For me it was a reminder that although we can have our fun, we must never forget that the animals need us. There will always be work for us to do.

Q11. LESSONS/ADVICE

Keep your eyes open to the truth, no matter how painful it might be. Only by being a witness to the truth can you make a difference. Educate yourself. Do your homework. People will look for opportunities to discredit you. Don't give them any.

Walk the talk. People will hold you to a higher standard than those by which they themselves live. They will look for inconsistencies, because it's easier to dismiss you as a hypocrite than to self-examine with respect to your message.

Don't adopt an "us versus them" attitude. Don't forget that virtually all of us were on the other side of the fence at one time. Don't be arrogant or smug. Lead by positive example.

Don't ever give up. Elie Wiesel said, "Take sides. Neutrality helps the oppressor, never the victim. Silence encourages the tormentor, never the tormented." Be part of the solution, or you will be part of the problem.

Know that you are not alone. Not only are there others out there who feel as you do, some of the greatest human minds in recorded history have espoused those same values.

Consider every opportunity to be a once-in-a-lifetime occurrence. Seize every chance to educate others, but don't shove your message down people's throats. You may be planting a seed that will grow when its time comes.

Don't bother arguing with people who simply tell you to get a life. If you are an activist, you have a life. Save your energy for those who really want to discuss or debate the issues.

Keep balance in your life, or you'll go crazy. There have been times when I have felt paralyzed by what I have seen. You can't be effective if you can't function. Count your blessings!

Q12. ADDITIONAL INFORMATION

Animal suffering is a human problem. We have a godlike power over the rest of creation. We can crush the life out of other living beings. Why not instead use that power to support and promote life? To do so is even more godlike.

One of my favorite quotes is from the book *One*, by Richard Bach: "Character comes from following our highest sense of right, from trusting ideas without being sure they'll work. One challenge of our adventure on earth is to rise above dead systems—wars, religions, nations, destruction—to refuse to be a part of them, and express instead the highest selves we know how to be."

When you see something wrong, when you think, "something must be done about this problem," consider that the solution might start with you. As Raffi said, "We must become the change we need in the world." You might be surprised at where it will lead you.

Alan Herscovici

Q1. BIOGRAPHICAL PROFILE

I was born and have lived most of my half-century in Montreal, Canada, an island city almost a thousand miles inland on the St. Lawrence River. Ships from Europe couldn't advance further than the Lachine Rapids, just west of Montreal, when the city was founded in the seventeenth century. This is why Montreal became North America's most important fur-trading center. (Many cities across the continent began as fur-trading posts established by adventurers who set out from Montreal.) I enjoy the fact that my family's involvement with the fur trade continues a long Montreal tradition.

I have spent much of my last twenty-five years writing everything from magazine articles and documentaries to political thrillers. I also enjoy reading, good conversation, playing and listening to music, and walks with my dog in the woods.

Q2. BECOMING INVOLVED

To understand how I became a critic of the animal rights movement, we could begin with my grandfather Armand, my father's father. His family fled anti-Jewish pogroms in Romania when he was a young child and he was brought up in Paris, where he apprenticed as a fur craftsman. As a young man, in the early 1900s, he immigrated to Canada. Each morning he joined other workers in the streets of the Old Montreal fur district, with his fur-sewing machine on his shoulder, waiting to be called in by one of the company owners who needed an "operator" for a day's work.

Alan Herscovici. Photo credit: Michel Gravel

My grandfather was a skilled fur cutter and operator, and after several years he brought his parents and siblings to Canada and started a fur-manufacturing atelier of his own. My father later joined him in this business, and so I was brought up in the fascinating world of the fur trade. Despite the glamour image of fashion furs, this was (and remains) a labor-intensive and highly competitive craft industry. Like the children of hardworking immigrant families everywhere, I was encouraged to get a good education that would provide an easier life. I duly cobbled together several degrees in economics and political science. Unfortunately for my game plan (of moving swiftly into a lucrative professional career), however, this was the late 1960s—the era of anti–Vietnam War protests, New Left politics, Timothy Leary, and Woodstock. Like many of my generation during those years of activism, I faced police horses in antiwar demonstrations; choked on tear gas with students in Paris in May of 1968; and participated in a "strike" that closed down the political-science faculty at McGill University in 1969.

I emerged from those turbulent years as a writer, traveler, and, I suppose, freelance activist. In the seventies, I visited the mountains of Chiapas and filed newspaper stories about the suffering of Mexico's aboriginal peoples—a struggle that would become international news much later, in the 1990s. My longest trip took me to Ladakh, western Tibet, the first year that isolated mountain region was open to foreigners after the Sino-Indian wars. I began writing about China's campaign of cultural genocide against the Tibetan people, some twenty years before Hollywood made this a trendy cause.

Travel weary, I retreated for a stint of "back-to-the-land" living with an extraordinary band of creative and adventurous people on the islands of Canada's west coast before, in the late seventies, returning to Montreal where I worked briefly with my father in his fur company. This confirmed that I wasn't cut out for a business life, but my reintroduction to the fur trade focused my attention on an activist cause that struck close to home. The International Fund for Animal Welfare (IFAW), Greenpeace, and Brigitte Bardot had rocketed the East Coast seal hunt into the international media spotlight. The white coat seal pup became a symbol for concerns about nature. What bothered me was that the antihunt protesters were then (and are still, too often) cited as "ecologists." How could ecologists propose that synthetics (usually derived from nonrenewable petroleum resources) were a better choice than sustainably harvested, natural furs?

Seeking to answer this question led me to study the emerging "animal-rights" philosophy, a concept that was not yet very well known. The results of this research were published in *Second Nature: The Animal-Rights Controversy* (Canadian Broadcasting Corporation 1985; Stoddart 1991). *Second Nature* was one of the first serious critiques of animal rights ideas and campaigns. In this book I showed how the animal welfare and conservation (ecology) concerns that sparked the first protests against the Canadian seal hunt, in the 1960s were quickly addressed. Instead of celebrating the speed with which government authorities (and hunters) responded, however, the protest movement escalated. In fact, Greenpeace and Brigitte Bardot didn't arrive on the scene until the mid-seventies, after good regulations were already in place. Subtly, the focus of the campaigns shifted: Although protesters still claimed that seals were being treated cruelly and could be hunted into extinction, their real goal now was to convince people that we shouldn't kill seals at all. But if it is wrong to hunt seals, what right do we have to kill any animal? That was precisely the point: the seal-hunt protests had become the first major "animal rights" campaign.

Through the 1970s and 1980s, the seal hunt became a testing ground where activists honed their media and fund-raising skills. Animal rights became a new "protest industry" with multimillion dollar budgets, expense accounts, and quasi-celebrity status for its leaders. By the time the European ban on the import of white coat seal pup pelts came into force, in 1983, the animal rights movement was ready to take on larger targets: laboratory animal research, the fur trade, circuses, and animal agriculture. As these campaigns intensified in the late 1980s, journalists, educators, government authorities, and targeted industries were looking for advice about how to respond to these new challenges. Almost without realizing it, with my research and writing, seminars and communications consulting, I too became an activist of sorts.

Q3. IMPORTANT ISSUES

The strongest motivation for my "activism" on animal rights issues has come from the wonderful people I have met through my research and writing. These people have included Cree and Inuit hunters communities across northern Canada, Newfoundland fishermen and Louisiana bayou muskrat trappers, Alberta cattlemen, animal trainers, and medical researchers from many countries. I have been impressed by how knowledgeable and passionate these people are about their work, and by the warmth and enthusiasm with which they welcomed me into their worlds.

I will never forget the Inuit woman who cried after I brought her to an animal rights lecture, she was so hurt by the way hunting and trapping were portrayed. A western dairy farmer's daughter also cried as she told how it felt to be called "cruel" because of her father's work— her father, who trudged out to the barn, yet one more time, each cold winter night before he went to sleep to be sure "his girls" were comfortable. And then there was the young Newfoundland sealer who thanked me for my writing because "it was good to know that someone out there understands our way of life." And the Cree Indian trapper who read *Second Nature*, footnotes and all, through a long winter season alone on his northern trap line. Each such encounter has rekindled and reinforced my determination to help people understand the terrible contradictions of the "animal rights" philosophy.

A second motivating factor for me is the need to respond to the muddled ideas and damaging impact of the so-called animal rights movement. It is troubling that this movement pretends to be "progressive" and preaches compassion for animals, but has no qualms about attacking the livelihoods, cultures, and moral integrity of people they barely know. There is something very wrong about a movement that talks about "expanding our circle of moral concern," but is prepared to use the propaganda techniques of the bully—incomplete or misleading information, media dirty tricks, even threats and harassment—to demean and intimidate the farmers who feed us, the scientists searching for cures for diseases that cause so much suffering, and some of the last surviving aboriginal hunting peoples on the planet. How can animal activists claim to care about "mother earth," but propose that we should wear petrochemical synthetics instead of natural wool, silk, fur and leather? The animal-rights doctrine is in complete opposition to the "sustainable development" principles of current environmental thinking as expressed by the World Conservation Strategy, the United Nations Environment Program, and every major conservation organization.

The third main motivation for my work is the moral arrogance of some of the animal rights leadership. I was shocked when Richard Adams (author of *Watership Down*) walked out of an interview with the *Boston Globe* rather than answer my explanations of why the fur trade is a well-regulated and environmentally responsible industry. (Adams was on a media tour, but apparently preferred not to share the platform with anyone who knew the issues well.) Most of the animal rights campaigners I have met have never visited a trap line or fur farm—but that hasn't stopped them from trying to destroy the livelihoods of people who live close to the land. Animal activists become very defensive when I suggest that they think they are morally superior to those they attack—but what other explanation is there for their refusal to respect the views of those who use animals?

The Animal Liberation Front (ALF) expresses a particularly dangerous form of moral arrogance when they destroy the property and life work of farmers, scientists, and others who don't share their views. This is an attack on the very heart of liberal democracy, and it is disturbing that so few animal rights advocates have spoken out against such tactics. This is something that should worry everyone who believes in the free exchange of ideas.

Q4. GOALS & STRATEGIES

[No response.]

Q5. DEFINING SUCCESS

My goal is to encourage some clearer thinking on animal rights issues. Much of my work has focused on explaining the difference between animal *welfare* (preventing unnecessary suffering of animals we use) and animal *rights* (the belief that people have no right to use animals at all, even for food or vital medical research). Although it may sometimes be difficult to know where to draw the line, the two concepts are fundamentally opposed. In fact, very few people in our society would endorse the strict animal *rights* philosophy, as expressed by writers such as Tom Regan (*The Case for Animal Rights*). This is why animal rights activists like to muddy the waters by playing up specific problems (real or fabricated) while avoiding serious discussion about the impact their campaigns have on other people.

I am very concerned that the current confusion about animal rights issues is aggravated by the nature of the communications media in our society. Animal rights lobby groups such as People for the Ethical Treatment of Animals exploit the media's weakness for confrontation and street theater. Placard-waving activists protesting at a busy city intersection are considered "news"; hard-working farmers who care for their land and animals (and produce food for our tables) are not. PETA unabashedly provides a "politically correct" pretext for publishing images of scantily clad young women; serious explanations by research scientists or wildlife biologists just don't have the same media "punch." PETA's stunts certainly generate media coverage, but they do little to encourage intelligent discussion of serious issues. If animal rights activists are so eager to use sexist images and other ad-man's tricks, how can they claim any moral credibility? Now that groups like PETA have become a new multimillion-dollar "protest industry," I believe it is time for them to be held accountable for their actions, just like any other industry.

I think that the animal rights debate is often more about symbols than reality. The white coat seal pup became a symbol for concerns about threats to our natural environment, even though harp seals are not endangered and the hunt was never an environmental threat. Anti-fur campaigns are often (especially in Britain) thinly veiled exercises in "class war" even though the real victims, were such campaigns to succeed, would not be the rich (who would wear something else) but down-to-earth working folks: aboriginal and other trappers, farmers and craftspeople. Campaigns against "factory farms" and research labs often play on our fears about globalization and rapidly changing technologies, rather than honestly addressing real animal-welfare issues. (Old Macdonald's farm provided little protection from heat, cold, drought, disease or parasites.) This confusion between symbols and reality explains why debates between animal activists and representatives from industry or science are so rarely productive. The two sides are speaking (more often, yelling) at cross-purposes.

For a writer who takes these issues seriously, all this can be very frustrating. Sometimes it seems that it would be simpler to take PETA's approach and deal with this cynically, as a public relations battle. Most days, however, I still like to believe that serious ideas and discussion can prevail in the end, even if this is a slower and more difficult route.

Q6. GROUPS & ORGANIZATIONS

My involvement with formal groups has changed and evolved over the twenty years that I have worked on animal rights issues.

For many years, I worked completely on my own, as a freelance writer. This is when I wrote *Second Nature*, produced a series of radio documentaries for the Canadian Broadcasting Corporation, and wrote many newspaper and magazine articles on these issues. I also worked as a consultant for the Canadian Royal Commission on Sealing, and participated in many interviews and debates about animal rights.

My research made me realize that the biggest problem with the animal rights controversy was that most of the people attacked by animal rights activists (sealers, fur trappers, farmers, research scientists) were not very good communicators—or had little access to the media. In the late 1980s, I began to work more directly with these people, encouraging them to tell their story. As a freelance communications consultant, I helped to produce educational materials for a wide range of agriculture, scientific, and fur-trade groups.

Three years ago, I was asked to take over direction of the Fur Council of Canada, the national, nonprofit association representing people in all sectors of the Canadian fur trade—some eighty thousand trappers, farmers, processors, designers, craftspeople, and retail furriers. An important part of my work with the Fur Council has been to produce a series of new educational materials that explain the environmental ethic of the fur trade. Our video (*Fur—The Fabric of a Nation*) and various printed materials have been requested by thousands of schools across Canada and abroad, and were promoted by UNESCO's environmental education program.

I have also maintained my contacts with many other groups involved with the responsible use of animals. I recently helped to test a new video (*Waiting at the Edge*) produced by the Arctic Inuit seal hunters of Nunavut, Canada's newest territory. It has been fascinating to watch the reaction of high school and university students who saw this video. Typical comments included: "All we ever heard about seal-hunting was the protests, so we assumed something was wrong . . . When you see that seals are not endangered and that the Inuit depend on this hunt for food and income, you realize that this is important for their culture . . . This is a side of the story we've never seen before." I heard the same reactions when a group of Inuit hunters spoke at the National Animal Interest Alliance meeting in Portland, last spring. Most people, even in this well-informed group, were shocked and dismayed to learn that the Marine Mammal Protection Act arbitrarily bans the import of Inuit seal products into the United States.

Reactions like these restore my faith that the extremist notion of "animal rights" promoted by sensationalist groups such as PETA will be consigned to the dustbin of history, as farmers, hunters, scientists, and others involved in the responsible use of animals begin to tell their own stories.

Q7. "THE OPPOSITION"

Any person or group that focuses on "winning" instead of truth, tolerance, and respect for others. Some people now use "animal rights" as a politically correct smokescreen to justify and incite intolerance of other people and cultures.

The concept of "opposition" is not that useful, however, because most people who support the animal rights cause are well meaning. I'm not so sure about the integrity of the leadership of wealthy animal rights groups like PETA, the Humane Society of the United States, and the International Fund for Animal Welfare. One rarely discussed advantage of positioning yourself as an "advocate for animals" is that those you claim to represent cannot challenge your leadership. In fact, most of the wealthy animal rights groups are remarkably lacking in basic democratic procedures even for their human members. Although groups such as PETA or IFAW claim hundreds of thousands of "members" or "supporters," they don't seem to feel any need for general membership meetings or elections.

However, I think PETA is a symptom, rather than the real problem. As the Dalai Lama says, "My enemy is my teacher." Animal rights campaigns reflect the fact that, for the first time in human history, the great majority of people in western society now live in cities, with little or no contact with the land. The only animals most people encounter are "pets" or tame squirrels or pigeons in the park. Many people take our abundant supply of nutritious food for granted. Few understand the role animal research has played in important medical advances. (Even animal rights supporter Paul McCartney recently admitted that he was surprised to learn, during Linda's illness, about the importance of animal research.) We need to encourage much more serious discussion of these important issues.

Q8. HEROES

I admire and am inspired by writers, scientists, and people of any profession who have tried to improve society and promote human welfare.

Q9. RELIGION/SPIRITUALITY

Even though I am not very interested in organized religion, certain of the ideals of my Jewish heritage have certainly influenced me. One important Jewish tenet is that there is no point seeking spiritual goals unless we have been honorable in our day-to-day dealings with other people.

I have also found spiritual and ethical guidance in the course of my advocacy work. Tibetan Buddhism, for example, teaches that compassion without knowledge is incomplete and can even cause great harm. Compassion and knowledge must work together if we are to fulfill our true potential.

In *Second Nature*, I devoted a chapter to explaining the spiritual view of hunting that underlies Native American Indian culture—the reciprocity between humans and the animals we depend on for our survival. This probably comes closest to my own feelings about the mystery of life and death in which we all are participants.

Q10. FUNNY/STRANGE EXPERIENCES

When I was invited to England to speak about animal rights, my hosts used special security precautions, because of concerns about the violence of the Animal Liberation Front. My name was registered with the front desk only as "Incognito." I didn't think anything of it until that evening, when I telephoned for a morning wake-up call. The receptionist confirmed the time I had requested, and then asked: "Is there anything else, Mr. Incognito?" I assumed she was just being ironic, until she added: "By the way, is your name Italian?"

Q11. LESSONS/ADVICE

Follow your interests and your beliefs with all the passion you can muster, but let your passion be tempered by reason—and always try to see the other side.

Q12. ADDITIONAL INFORMATION

[No response.]

Alex Hershaft

Q1. BIOGRAPHICAL INFORMATION

My name is Alex Hershaft. I was born in Warsaw, Poland, and I am a survivor of the infamous Warsaw Ghetto. After the war, I spent five years in an Italian refugee camp, then emigrated to the United States from Italy in 1951.

I received a B.A. from the University of Connecticut and a Ph.D. in Chemistry from Iowa State University. In the subsequent twenty-five years, I have done teaching, materials and operations research, and environmental management for academic, aerospace, and consulting institutions.

I enjoy humor, music, dancing, sports, and righting wrongs. I also enjoy seeing my daughter work; she is a professional actress.

Q2. BECOMING INVOLVED

I was always bothered by the idea of hitting a beautiful, living, innocent animal over the head, cutting him up into pieces, then shoving the pieces into my mouth. I finally made my decision to stop eating animals when I came upon a ritual slaughter scene during a visit to Israel.

My experiences in the Warsaw Ghetto during the Nazi Holocaust had a profound impact on my subsequent life choices. I felt some guilt that I lived when so many others didn't, and a sense of duty to redeem my survival by assuming their share of responsibility for making this planet a better place to live for all its inhabitants. After the war, I became active in the religious freedom, civil rights, peace, and environmental

Alex Hershaft.

movements, receiving much fulfillment, but always feeling that I was missing something. In the summer of 1975, I attended the World Vegetarian Congress in Orono, Maine, and had a deeply emotional experience. Subsequently, I took time to reflect on the root of the key problems challenging planetary survival, that is, disease, hunger, environmental devastation, oppression, and war. Amazingly, all evidence pointed to animal agriculture as the common root cause. My life's mission then became crystal clear. In particular, my experiences in the Nazi Holocaust allowed me to empathize with the condition of farm animals in today's factory farms, auction yards, and slaughterhouses. I know firsthand what it's like to be treated like a worthless object, to be hunted by the killers of my family and friends, to wonder each day if I will see the next sunrise, to be crammed in a cattle car on the way to slaughter.

Q3. IMPORTANT ISSUES

Opening the eyes, the minds, and the hearts of the American people to the evil of eating animals. I care about this issue because it goes to the core of all I believe in: justice, public health, world hunger, and environmental quality. Other people should care about this because it defines who they are. Animal rights is not so much about "them" as it is about us.

Q4. GOALS & STRATEGIES

My strategy is to change people's eating habits by informing them of the evils of eating animals (and their products), and by making meatless foods readily available and desirable. We accomplish this primarily through national grassroots campaigns, such as the Great American Meatout, and also through placement of hundreds of letters to the editor. In addition, I have personally written numerous articles and lectured extensively on animal agriculture, farm animal abuse, diet and health, and environmental conservation. I also lead intensive seminars on personal growth, persuasion, and campaigning.

Some cynics have questioned the effectiveness of the vegetarian and animal rights movements. They note that U.S. and world meat consumption is rising, that factory farms are getting bigger and more oppressive, and that the Humane Slaughter Act is a sham.

Yet the problem is not that we're not effective, but that we fail to recognize the magnitude of our challenge, the nature of social struggle, the proper strategy to pursue, and the nature of winning.

To recognize the magnitude of our challenge, consider that flesh eating is more deeply ingrained in our social fabric than smoking or the exploitation of women and minorities. Consider also how many decades it has taken to work on these social ills; for animal rights, we have been at it less than twenty years.

Each struggle for social justice goes through at least three phases. In the Alerting Phase, we call public attention to the problem through outrageous acts. In the Discussion Phase, we appeal to the feelings and beliefs of those whose attention we have won. In the Reform Phase, we implement needed reforms through legislative, administrative, economic, and social pressures.

In the Alerting Phase of our struggle for plant-based eating and farmed animal liberation, we blocked USDA and slaughterhouse gates, occupied the office of the secretary of agriculture, dressed in animal costumes, and held countless pickets and vigils. That was over by the mid 1980s.

The strong public disapproval of farmed animal abuse, promotion of plant-based eating by mainstream health advocacy organizations such as the American Cancer Society and National Cancer Institute, and wide availability of meat and dairy alternatives provide ample evidence that we are now in the advanced stages of the Discussion Phase. This is when we should befriend our targets, appeal to their feelings and beliefs, and stop their subsidizing of animal agriculture at the supermarket checkout counter. Yet a leading national organization insists on impeding our progress by persisting in Alerting Phase tactics.

The transition to plant-based eating will continue gradually into the Reform Phase, when consumption of animal products will be relegated to special venues and special occasions, as it is in most of the world today. It will receive occasional boosts from outbreaks of meat-borne diseases, like the "Mad Cow" disease. The transition in the treatment of farmed animals will also be driven by trigger events, like slaughterhouse exposés or regulatory reforms by large fast-food chains.

Finally, we need to recognize that real-life winning in the Reform Phase has a vastly different flavor from the utopian visions we once entertained.

People will not be joining vegetarian societies en masse—they will just eat more fake meat and dairy products, and they are. Mainstream health institutions will not be replaced by vegetarian societies—they will just promote plant-based eating, and they are. Supermarkets will not be replaced by vegetarian health food stores—they will just carry more meat and dairy analogs, and they are. McDonald's and other fast-food chains will not go out of business—they will just push veggie burgers and humane reforms, and some are. Slaughterhouses will not be reduced to smoldering ruins—they will just produce meatless foods, and some are.

Yes, we are winning! But, we need to be persistent, realistic about our victories, and sensitive to the long-term impact of our actions on mainstream public opinion, rather than the fleeting impact of our personal fulfillment.

Q5. DEFINING SUCCESS

Success is when vegetables and grains are considered the main staple of the American diet, and meat is generally used as a condiment.

Q6. GROUPS & ORGANIZATIONS

In the summer of 1981, I organized "Action for Life," a national conference that launched the U.S. animal rights movement. That same year, I gave up a successful career in environmental management and an affluent suburban lifestyle to devote my full attention to exposing and ending animal abuse and other destructive impacts of animal agriculture.

Later that year, I founded FARM (Farm Animal Reform Movement), which has become a major force in the struggle for vegetarianism and improved treatment of farm animals. FARM holds annual campaigns such as the World Farm Animals Day and the Great American Meatout. In addition, FARM sponsors National Veal Ban Action, Letters from FARM, CHOICE (of plant-based meals in schools), and several other programs. FARM has held many national conferences which have turned hundreds of concerned Americans into animal rights activists.

In all these efforts, we have attempted to work closely with other organizations and to submerge our personal and tactical differences in the common pursuit of animal liberation from all forms of human oppression.

Q7. "THE OPPOSITION"

All groups and individuals who profit from the suffering and death of animals are the opposition.

We advocate a tax on meat and other animal products. Part of the proceeds should be devoted to retraining people in the meat industry into socially useful occupations.

Q8. HEROES

Mahatma Gandhi, Henry Spira, and Howard Lyman. I also believe that each of us should conduct him/herself in such a manner as to serve as a role model to others.

Q9. RELIGION/SPIRITUALITY

What drives me is a deep respect for life and our natural environment and an abhorrence of injustice. If you wish to call that a religion, you may.

Q10. FUNNY/STRANGE EXPERIENCES

Almost all the civil disobedience activities I have engaged in appear a bit ludicrous and incongruous from hindsight, but then they had to be to bring us favorable attention.

Q11. LESSONS/ADVICE

My advice is to select the most effective organization, to plunge in body and soul, but to take the time to "smell the flowers." Beyond that, to focus on actions (good or bad), rather than the people taking those actions.

When we tell someone that a vegan diet *can be* nutritionally adequate or preface an affirmation of our compassion for animals with "this may sound crazy, but . . . ," we invite rejection of our position. When we put up billboards asking "Got Beer?" we invite association with beer-chugging, drunk-driving hooligans. When we torch a slaughterhouse, we invite an image of us as "terrorists," no different than the men who blew up the U.S.S. *Cole* in Yemen. These are some elementary examples of "positioning" our movement for certain failure in the court of public opinion.

We need to get over our apologetic attitude by taking time to learn the arguments and by realizing that people are now open to our views and will respect us for presenting them in a calm, logical, caring manner.

We need to temper our desire for short-term media attention with the long-term consequences of our means of getting the attention.

Q12. ADDITIONAL INFORMATION

The term "vegan" defines for me a lifestyle that is as free of cruelty as is practical, without substantially detracting from my life's mission. It means choosing foods that are free of animal flesh, dairy products, eggs, honey, and ingredients derived from these products (e.g., gelatin, casein). It means choosing apparel free of leather, fur, wool, silk, and other products of animal exploitation. It means choosing personal care, household, and other products that are not tested on animals. It means making the more difficult choices to use gelatin-based photographic film, rubber tires that may contain animal ingredients, and, occasionally, foods containing sugar that may have been refined on charred animal bones. It does not mean prowling through my friends' refrigerators and medicine cabinets in gleeful search of flawed items and making their lives miserable. In short, it means treading gently and lovingly in an imperfect world.

J.R. Hyland

Q1. BIOGRAPHICAL PROFILE

I was born at the beginning of the baby-boom generation, and like many other people in the postwar era, my parents moved from the New York City area to the suburbs of Long Island. My father was a journalist, as his father had been, and my mother was a homemaker who was glad to have escaped the clerical job she had before she married. Both parents assumed my career would be marriage. Unlike many young girls, I never spent any time daydreaming about myself as a bride—or as a wife or mother. But neither did I ever daydream about an alternative. So at nineteen I married, but was widowed when I was twenty-eight. I never remarried and although I had been raised in a Roman Catholic home, and attended Catholic schools, I enrolled at a Bible college/seminary and was eventually ordained as an Evangelical minister.

Q2. BECOMING INVOLVED

After my husband died, I went back to school, this time to a liberal arts college. My original studies in biblical theology fit me only for teaching or pastoral work, and I didn't want to do either of those things. So I decided to major in psychology and get the credentials I would need for counseling. But then my studies brought me in contact with the psychology lab. Rabbits, rats, cats, and dogs were subject to atrocious experiments from which nothing was learned: they were merely replications of existing studies. The only thing being accomplished was a desensitization of those who were able to stifle the compassion that would interfere with their "objectivity."

J.R. Hyland. Photo credit: Jean Burns

This encounter took place in the early 1970s, and there was no animal rights movement; no support for the belief that it was insensitivity, not objectivity, that allowed one to continue on with the gratuitous torment of other creatures. And, certainly, back then, there was no latitude that allowed you to avoid taking courses that violated your sense of ethics or morality. Neither was there any support or concern for the predicament of the animals. The consensus of the professors and other students to whom I voiced my concerns was that I was a crank making a big deal out of nothing. They said, in effect, "Shut up, take the courses, get on with it and get your credentials." But I couldn't do that, so I changed my major. I ended up in the Philosophy department, eventually doing undergraduate and graduate work in the division of religious studies.

For quite a few years, my experience with animals was limited to a concern for stray cats and dogs that came across my path and for the companion animals who were considered part of our family. But this caring did not seem to have anything explicit to do with my religious beliefs. Then, in 1985, I was asked to contribute a paper to an ecumenical conference. I was to show the ways in which the concept of animal sacrifice had impacted both Judaism and Christianity. Although I disliked the actual process of writing, I liked doing the necessary research: it was like detective work, often leading to unexpected conclusions. I couldn't remember any dialectic on the subject of animal sacrifice from my seminary days, but now it was the decade of the 1980s and animal welfare was being discussed in a new way. I was pleased: not only was my presentation well-researched, but it was also timely. Unfortunately, I didn't get to present my scholarly, timely paper to my Christian colleagues. Not then. Not ever. When it was reviewed, the paper was summarily dismissed: my research had led to an unacceptable conclusion.

I had found that from Genesis to Revelation the Bible bears witness to a story of the human failure to fulfill the role that God assigned—the role of compassionate caregiver for other species. Just as our Western culture used the Bible to denigrate women, to prolong slavery, and to justify the violence of war, the Scriptures had also been used to establish a reign of terror against animals. These other sentient beings had been tortured, maimed, and killed with a self-righteousness that claimed divine sanction for its destruction of the other creatures with whom we share the earth. And animal sacrifice was part of that self-serving process. But no one wanted to hear this.

I felt as though I were living in the midst of the story of "The Emperor's New Clothes." The king had accepted the lie that he was beautifully dressed and that only the righteous could fully appreciate the beauty of his raiment. The entire adult community considered themselves righteous so they, too, said he was beautifully dressed. However, far from being clothed in beauty, their leader was walking around nude. But a collective consciousness had been established on a lie, so consensus overrode reality. The same kind of consensus had triumphed over the reality of the biblical witness regarding animals. And now I wanted to find out what others had to say about this ongoing abuse of animals, but I couldn't find any books on the subject.

So I began writing *The Slaughter of Terrified Beasts: A Biblical Basis for the Humane Treatment of Animals*. It was published in 1988.

Q3. IMPORTANT ISSUES

Until the publication of *The Slaughter of Terrified Beasts*, I was so involved with the work I was doing with migrant farm workers, and in prison ministry, that I had no idea of the extent of the animal rights movement. But then I started hearing from many of my readers and was very glad to find that vegetarianism was an important issue.

Although I did not eat animals, I had believed it would take another hundred years before this could become a viable movement.

And I knew that as long as people insisted that other creatures existed to be killed and consumed by them, their consciousness would be closed to many other forms of animal abuse. This was especially obvious to me in the attitude toward recreational killing—hunting.

Many who oppose those who maim and kill animals for "sport" will support those same weekend hunters who insist that they will eat the animals they have killed.

Whether it is recreational killing, human carnivorism, or the many other ways in which we torment and destroy other creatures, these cruelties are reflected in the way humans treat each other.

When someone is terribly abused, we are outraged, saying they did not deserve to be "treated like animals." People, like animals, have been hunted, tortured, vivisected, starved, beaten, shot, and stabbed. Like animals, they have been burned to death, sexually abused, skinned alive, and eaten. Human abuse and animal abuse are two sides of the same coin: you can't have one without the other.

Q4. GOALS & STRATEGIES

Although my goal is to provide a biblical basis that will help to bring about changes in traditional Christian attitudes toward animals, this is not something that can be directly accomplished. There is enormous resistance on the part of priests, ministers, and various church groups to the biblical teaching that both humans and animals are "nefesh chaya": beings who live because of the soul infused by their Creator. Consequently, we provide support and encouragement to nonbelievers, as well as to those people of faith who are part of the spiritual evolution manifesting itself among those who understand that kindness and compassion must be the hallmark of our relationships with all creatures.

Under the ministry name of Humane Religion, we publish and distribute a variety of materials dealing with issues such as hunting, vegetarianism, and the wearing of fur. We also develop humane education materials and function as a resource for individuals and for various groups—both secular and religious—helping them to develop seminars, groups studies, materials for Sunday School classes, or whatever other needs come to our attention. And for several years we published a bimonthly journal, but the cost of printing and postage became prohibitive and we had to stop publication. However, many of the articles printed in the journal are available on-line at our Web site. And we have been a resource for an ongoing court case in which Christian inmates petitioned the courts for the right to a vegetarian diet, in accord with their religious beliefs (Muslim and Jewish prisoners are already allowed special diets). Basically, we do the necessary research and offer whatever other support we can give, as the need arises. There is no charge for these services.

Q5. DEFINING SUCCESS

My goal is to have been part of a spiritual evolution that will ultimately bring about a way of life which approximates the world spoken of by the Prophets of Israel and by Christ. They told of a world in which all species, human and animal, will live together in the kind of Peaceable Kingdom of which Isaiah spoke. It would be a time and place in which "nation will not take up sword against nations, nor will they train for war anymore." This kind of world would reflect the goodness and love of the heavenly kingdom for which Christ taught us to work and to pray; a world in which "[God's] will be done, on earth as in heaven."

Q6. GROUPS & ORGANIZATIONS

No, we don't work closely with any particular group, but we do share our resources and cooperate with many animal rights organizations from large groups such as People for the Ethical Treatment of Animals (PETA) and the Fund for Animals, to smaller single-issue groups that are working to end the horror of things like the fur trade and factory farming.

Q7. "THE OPPOSITION"

There are neither groups nor individuals that I think about in that way. Because no matter how great our understanding, or how lofty our goals, we all have blind spots that cause us to be part of the problem and, in some way, obstacles to the coming of the Peaceable Kingdom. I have known vegetarians who do not eat flesh because of the killing involved, but they are militant nationalists. I have known peace activists who refuse to acknowledge the violence involved in the killing and eating of animals. I have known those who work hard for the welfare of cats and dogs, but think nothing of wearing the fur coats that are the graves of animals who were killed. The New Testament puts it succinctly: "All have sinned and fallen short of the glory of God."

Q8. HEROES

From the time I first read their oracles, the Latter Prophets of Israel have been my heroes. They were great and godly men who understood that the worship of the Creator entailed something other than ritual observances and prayers. Men such as Amos, Isaiah, Hosea, Micah, and Jeremiah called for a religion of social justice: a religion in which the worship of the Lord was expressed in the concern and the care of the poor and the powerless. They also spoke out against the abuse of other species that was ritualized in the practice of animal sacrifice. And in modern times, I am encouraged by the work of such men as William Wilberforce and Lord Shaftesbury whose Christianity was an active force that impelled them to successfully work for the abolition of the slave trade in England and to initiate programs to prevent cruelty to animals and the oppression of those powerless humans who also suffered at the hands of a callous and nominally Christian society.

Q9. RELIGION/SPIRITUALITY

When I was twenty I had an ongoing, spiritual experience that continued for almost a year and changed my life. Although it had nothing to do with traditional religion, it did lead to my making a commitment to try and live my life in accord with the Gospel teachings of Christ. I understood his ministry to have been a continuation and expansion of the principles of social justice, compassion, and nonviolence contained in the oracles of the Hebrew prophets. I followed this commitment to the best of my ability, working in various para-church ministries through the years (para-church indicates something other than pastoral ministry in a church setting). After 1988 I was increasingly involved in animal rights activities, but my primary work continued to be with migrant farm workers and in prison ministry. Then in 1995, after much inner prompting, I was able to make the work for animals my primary commitment. At the time I didn't really know what that meant—there were no established ways for a minister to do this. It was only after the commitment was made that I began to find ways to fulfill it.

Q10. FUNNY/STRANGE EXPERIENCES

[No response.]

Q11. LESSONS/ADVICE

I would caution those who are just beginning to work for animal rights not to be discouraged by the strong differences of opinion they may find. This work needs people who are passionately committed and caring, and their intolerance for a barbaric status quo can sometimes become an intolerance for the way others are trying to end animal abuse. And, occasionally, the tactics of one group will be so unacceptable that another group becomes afraid it will set back the struggle for animal rights. But it won't. There are so many cruelties that must be ended, and so many ways of providing care to be instituted, that it will take all kinds of activism to bring this about. The radical, the middle-of-the- road, and the conservative among us all have a contribution to make.

Q12. ADDITIONAL INFORMATION

I think it is important to understand that those who are aware of the cruelty involved in the treatment of other species are morally bound to take some kind of action to end it. Kindness to those animals who come across our path is not enough. The needs of those who are hidden from our sight, enduring the atrocities inflicted on them by factory farming, by research labs and by "sportsmen" are just as important as the needs of the companion animals with whom we share our lives.

Many people do good things, but too few go on to resist evil. Being kind to humans or to animals usually brings approbation and support, while resisting evil brings criticism and opposition. Nevertheless, along with the good that we do there must also be some kind of active, nonviolent resistance to an abusive status quo. That resistance can take the form of ongoing letters to newspapers, community leaders or political representatives, urging the end of specific cruelties. It can mean taking part in a planned protest or supporting those individuals or groups dedicated to ending animal abuse. What we are able to do is not as important as our willingness to take whatever action our lifestyle allows. Each of us is morally responsible for the kind of stewardship that human beings exercise over the earth and over the other creatures with whom we share this planet.

Roberta Kalechofsky

Q1. BIOGRAPHICAL PROFILE

I was born in Brooklyn, New York, in 1931. I have a B.A., an M.A., and a Ph.D., all in English literature. I am a novelist and a publisher. I became a publisher in 1975 with the intention of publishing fiction, never dreaming that I was going to wander into publishing animal rights and vegetarian books. You might say that the animal rights movement made a rather large detour in my writing life, and it is an effort for me to manage both sides of my life. However, I grew up in the small press movement as a publisher. Looking back over the last twenty-five years, I realize that I matured in grassroots movements and that that has been an extraordinary place to be. It has taught me much about the political system.

There is a "fun" side to my life. I do much exercising, aerobics, weight lifting, walking, and so on, love parties, love to dance, love beautiful gardens, regret that I don't have one; love to read, to visit old friends, regret that I don't have enough time to do these things. I could use three lives.

Q2. BECOMING INVOLVED

It was several "single" events. Richard Schwartz sent me a copy of his manuscript, *Judaism and Vegetarianism*, around 1982 or 1983. For the first time, I read a description of factory farming, and it revolted me. I stopped eating meat, and eventually published his book. Almost simultaneously, by happenstance, I read a novel, called *Skin* by the Italian writer, Curzio Malaparte. In it there is a horrific description of a

Roberta Kalechofsky. Photo credit: © Audrey Gottlieb

man who loses his dog and finds him in a research laboratory. The chapter of this action is called "The Black Wind." My life changed instantaneously. I felt as if a terrible vision had been revealed to me. I sucked in my breath and have hardly been able to let go since. Combined with reading the description of factory farming in Richard Schwartz's book, I realized that the evil visited upon animals is beyond calculation. I wrote a prose poem called *The Sixth Day of Creation*, incorporating "The Dark Wind."

Discovering this terrible terrain of cruelty led to my founding Jews for Animal Rights, which I incorporated with my publishing company. Micah Publications became the publishing arm of Jews for Animal Rights. You can find information about both, on my Web site, *www.micahbooks.com*. The process of how I became involved in the animal rights movement has been described in my book, *Autobiography of a Revolutionary*. The book takes its title from the first essay, which was published in the journal, *Between the Species*.

Q3. IMPORTANT ISSUES

Animal agriculture and animal research are the two issues on which I spend my writing and publishing time. There are many other important issues in the animal rights movement, but those are the two that I concentrate on. I think the importance of animal agriculture is obvious. The Union of Concerned Scientists issued a statement that animal agriculture is the second largest cause of environmental decay. Almost daily now we read in the papers of the widespread problem of food poisoning. The dangerous bacteria *E. coli* 0157:H7 is now estimated to be very widespread in cattle. Food is basic to life, and it is being contaminated because of our agricultural policies.

The pain visited upon the animals in this decayed process is beyond telling. When I first encountered Harriet Schliefer's statement "Meat Is Murder," I thought it was histrionic. I'm afraid it is rather apt for both humans and animals. The Roman poet, Plutarch, who was a vegetarian, wrote, "What a world of pain we create for a little taste upon the tongue." What is sad about the cruelty visited upon animals in the agricultural system is that it is unnecessary. "Just say no." Every one of us has the power to eliminate enormous pain just by keeping our mouths shut.

As for animal research, I believe it is satanic. I realize that sounds melodramatic, but I don't use that word lightly. Some things like concentration camps and modern warfare are satanic. Laboratories are places that are as fiendish as slaughterhouses. What makes the problem worse in the case of animal research is that the people involved are presumably intellectuals—intellectuals without moral concern or moral awareness. The animal research laboratory is an excellent example of the derangement and fragmentation of modern intellectual life—the split between intellect and ethics. The poet, John Cowper Powys, said of vivisection: "To torture for knowledge—only the devil could invent such an idea." Animal vivisection led to and leads to experimentation on human beings, which is a great human rights problem that is overlooked. There has been and is much experimentation on human beings. It was not confined to Nazi Germany and the Japanese prison camps, but was and is practiced in hospitals in many western countries that practice western medicine. You can read several articles I have written on this subject on my Web site. Scroll down to "The Reading Room," and click on the subject you want to read about. You will also find essays there on Hitler's vegetarianism (only partly) and on whether the Nazis practiced vivisection on animals. (Of course, they did.) I receive so many questions on these two subjects I decided to post articles about them on my Web site.

I also think animal research, as a medical strategy, is misguided. It has borne some fruit, but fruit that is out of proportion to the harm it has done and the pain it has caused to both animals and humans. Medicine became reductionistic in the nineteenth century and turned away from the wisdom of prevention, the study of nutrition and herbs was downgraded, and the path from the animal laboratory has led to an overemphasis on drugs rather than on environmental balance. This emphasis has trained the public to expect "quick fixes" and to surrender individual responsibility for their health. We are rapidly approaching an era when the pitfalls of an overmedicated society are becoming obvious. We do research on primates for AIDS and cancer, yet more people die of staph and other bacterial infections in the hospitals, because doctors and nurses don't adequately wash their hands. An article in the *New York Times* (November 9, 1999) states that twenty thousand people a year die "as a direct result" of ignoring this simple procedure; "by contrast 17,171 Americans died of AIDS in 1998. And hospital-acquired infections will contribute to the deaths of 70,000 more people, far more than the 44,190 Americans who died of breast cancer in 1997." Semmelweiss died for nothing! Doctors could save lives just by washing their hands, but there is no technological satisfaction in that for researchers, no Nobel prizes, no grant money. Doctors could save thousands of lives by counseling heart patients about diet or by protesting the use of antibiotics in animals. There are many sane, rational, benign, noncruel ways they could save lives.

Q4. GOALS & STRATEGIES

I don't have time to do everything I would like. As a writer and publisher I concentrate on writing and publishing.

Q5. DEFINING SUCCESS

When everyone is a vegetarian; no one wears fur or leather; animals aren't used to entertain people in zoos and circuses; when the cages in zoos and research laboratories have been abolished; and when we realize that animals (like human beings) have been created to experience the joy of life, and pursue life, liberty and happiness—as they perceive it. In other words, when they are free, as we wish to be free ourselves. Cages are terrible things for animals or for people.

Those are my long-range goals. My short-range goals are to keep plugging in this direction, work for organic food, reform of farming and agricultural practices, reform in medical practice, work on environmental issues—the usual list of reforms our civilization needs now. I also take an interest in other abuses, such as slavery in the modern world. The antislavery society has a Web site that people should know: *www.antislavery.org*. It pleases me to remember that the animal rights movement, the abolitionist movement, and the women's movement all began about the same time in the nineteenth century, and that many of the founders of the antislavery movement were animal rights people.

Q6. GROUPS & ORGANIZATIONS

I maintain membership in all the usual standard animal rights and vegetarian organizations.

Q7. "THE OPPOSITION"

That would be a very large part of my society, as my point of view and the point of view of others like me is a distinctly minority viewpoint. In the nineteenth century, animal research was denounced by almost every major writer, but now I must accept my minority position, and hope and believe that it is part of a historical process that involves change. The "opposition" is part of that process too. Rather than think of anyone as the "opposition," I prefer to think that a younger generation will emerge with ideas closer to my own and that of past generations, which felt a greater reverence for life. Every "establishment" is inevitably the "opposition" to the generation that is working for change.

Q8. HEROES

There are many heroes and heroines in the movement. Other than my husband, who has been my right hand in this work, I don't wish to name names because that might invite invidious distinctions. I draw ideas and willpower from a compelling array of people, philosophers, lawyers, doctors, intellectuals, and nonintellectuals, people who do outreach, carry banners in rain and cold, march with their children, people who have given up careers to work in this movement. The animal rights movement is an extraordinary altruistic movement. People give up time and money because they are sick to heart at the cruelty inflicted on animals, and they do it without possibility of reward. There are no Nobel prizes for people in this movement, no medals, no votes, and no upwardly mobile careers. There is much talk today about the selfish "me" generation of the modern world. The animal rights movement marches in exactly the opposite direction of the "me" generation. It has put cruelty at the top of the list of the moral agenda.

Q9. RELIGION/SPIRITUALITY

It is a part of my work. The opening stanzas of Genesis and the Book of Job are foundation stones of my relationship to nature and the animal world, and I am not in the least bit hindered by my religious feelings in this regard. I believe that the prophets are on my side.

Q10. FUNNY/STRANGE EXPERIENCES

Yes, when I was writing *Haggadah for the Liberated Lamb*, which is a haggadah for a vegetarian seder. The Haggadah is the book Jews read on the night of the Passover, and the seder is the meal they eat. Friends of mine pleaded with me not to write it because they thought it was a waste of time. I remember saying to a friend, "If twelve people need this book, I must write it." In working out the gender problem about whether to refer to God as "He" or "He/She" (which I think is a linguistic abomination), I hit on the idea of referring to God as "You," and felt that I was in direct communication. It was a very intense experience. As the years have gone by, I have met dozens and dozens of Jews who have used that haggadah and have told me how much they loved it. Together with another vegetarian haggadah, *Haggadah for the Vegetarian Family*, we have sold over ten thousand copies. So I reached more than my original goal of "twelve people."

I also like to think that naming my publishing company "Micah" after the prophet was—if you will permit an unsubtle reaction—"prophetic." I like to think that a vision that is twenty-six hundred years old has ripened into a broadly based, transnational revolution in sentiment, which, as T.S. Eliot observed, is the only real revolution.

Q11. LESSONS/ADVICE

Don't spread yourself too thin. Concentrate on the one or two things you're best at or love to do. If taking care of pets is what you like to do, work in an animal shelter and/or on the pet industry. If you're housebound for one reason or another, but can do writing campaigns, do that. The movement is a vast moving front. You can plug into it almost anywhere: write, march, sign petitions, act in plays, write songs, get out into the schools, make films, and so on.

Don't let "the opposition" rile you, don't lose your good manners or your sense of humor. Don't worry about people who tell you that you're wasting your time, and don't waste your time on them. Don't worry about disappointments (there will be many), join support groups and celebrate victories with them.

Educate yourself. Know your subject. It's not enough to protest, you must protest knowledgeably. Clutch pleasures. Cultivate them. You may have dark moods, and you will need pleasures to sustain your balance.

As for the "opposition," read our material. Stop thinking of us as antiscience obscurantists or as anti-intellectuals or as new age weirdoes; we are morally serious people who can defend our positions. We know where you're coming from, because some of us have been there, and left. Do you know where we're coming from?

Q12. ADDITIONAL INFORMATION

[No response.]

Crystal Kendell

Q1. BIOGRAPHICAL PROFILE

My name is Crystal Kendell. I am currently a student at Utah State University studying Geology. I spent two years at the University of Utah as a Physics major, but decided that Geology was more to my liking. My father is a geologist, so I was raised on the stuff. He would bring home samples and hold me up so I could see them through his microscope. It was great.

I was born not too many years ago in Billings, Montana, 1979, on a cold November day. I moved to Utah when I was ten, and it was here that I discovered that using animals is morally wrong. I lived where a city was not too far away, but farms surrounded me. There was a small ranch nearby that specialized in training horses for show. There were many cattle, and even a large chicken factory down the road. Living in that type of setting really gives a person a unique perspective.

I enjoy many things, but chief among them is running. I ran track and cross-country all through high school and my first year of college. I've always been athletic, and enjoy being active.

I also love to do graphic design. That is one thing that I have been able to use as an activist: there is always a newsletter or flier that needs to be put together.

Q2. BECOMING INVOLVED

I became involved in animal rights issues slowly. I had been a vegetarian for a few years but felt I wasn't doing enough. I went through a transition to veganism, and then to activism. I didn't just want to be

Crystal Kendell. Photo credit: Myq Larsen

vegan myself: I wanted others to see that they could make that decision too, and that it's really not that hard, or unhealthy (or any of the other false ideas that people have about not eating animals).

It was no single event that started me down this path, but I must credit Sean, a friend, for pointing out the alternatives through the years that helped me make my pivotal decisions. He was always there with support and ideas. It really made a difference in my life.

Q3. IMPORTANT ISSUES

It's hard to say what issues I spend the most time on, because there really are so many. Still, the two that are special to me are the plight of farm animals, and the fur industry. I think that my interest in these is high because of where I live. There are many dairy cow feedlots and fur farms here. I drive past them, seeing them almost every day. I can see the cows with such huge swollen udders that they need "bras." I see where they spend their short lives. I see the mink in their small cages inside a shed, spinning their necks around and around from psychological stress (and inbreeding). It's all really open and right in front of our eyes, but not many people care to notice or attribute feelings to these animals. It is so convenient not to notice or care. Who wants to see atrocities when they can just brush them off? I ignored this for many years before my conscience got the better of me.

Q4. GOALS & STRATEGIES

My main goal will always be education. I'm always looking for new ways to get the word out: to get people to think about what they are doing and what they are eating. At my university, I am part of a group that offers information almost every week. We bring in speakers, hold discussions, write letters, and do all sorts of activities. I host a vegan cooking class once a month. This is a great way to actively show people how easy it is to cut out the meat and dairy products. Once I set up a tent on campus and lived in it for a month (along with a friend). It was to bring awareness about the animal research that is done on the campus. These strategies come and go, but the important thing is to have a presence, to be visible so that people can't ignore you or turn away from what is going on.

Q5. DEFINING SUCCESS

I think that success would be if everyone changed their thinking, just a little, to believe that animals are not our property. Success is not just having everyone become a vegetarian. It isn't the vegetarianism that is going to help animals, it's the mind-set. I have known many people who stop eating meat without changing their thinking. They almost always go back to meat consumption. They haven't changed their lifestyle, only their choice of what's for dinner. For an activist, it's never a "choice," it's just what you do because it is right. Success, I believe, will be some time coming, but it is not unreasonable to think that slowly people can and will accept animals as living, feeling creatures with a strong desire to survive. Once they think this way, they will no longer exploit the animals.

Q6. GROUPS & ORGANIZATIONS

I work closely with a couple of formal groups in my area. There is a college group at my university, and a few local community groups. Although we have few members at school, we have a strong and powerful voice. In the community there are two groups in particular. One is called the Utah Animal Rights Coalition, and the other is the Coalition to Abolish the Fur Trade, Utah chapter. Both are small, but active. It creates a great animal rights community where I can find people that are at least somewhat like-minded. It helps take away the enormity of what we must face. It does become frustrating sometimes.

Q7. "THE OPPOSITION"

There is no one group or type of person that I consider to be the opposition. It's hard to point at one person and say "that person is the enemy." We are not fighting a battle. This is not just a black and white issue; it is varying shades of gray. Occasionally the opposition can be a person: usually someone with a violent nature, but those types of people can be found in any community. Many believe ranchers to be the problem, but they are some of the nicest (though most stubborn) people I've ever met. They will listen and give feedback. After talking to some, we usually end up "agreeing to disagree," but they are seldom rude or confrontational. I often learn new things from these people.

I will cite the Fur Commission USA as a really difficult group to deal with, along with some of the local pig farms. But within these groups are people with a free will and minds of their own.

Q8. HEROES

I admire many people in the movement, but the two people I look up to the most are both personal friends. One happens to be my best friend, Daniel. He inspires me, motivates me, and he's not afraid to tell me if I'm acting ignorantly on a certain issue. The other, Eric, lives in a different city, but we went to college together for two years. He taught me about everything I know about being an activist. These two aren't famous by any means, but I look up to them more than I ever did some dead (or living) person whom I'll never meet.

Q9. RELIGION/SPIRITUALITY

I have never really been a religious person. I've often been questioned that if I don't have a faith, how can I claim to have any morals? These people seem to believe that if you don't have some type of fiery afterlife or supreme being to be afraid of, then you won't be honest. Other than that, my lack of spirituality has never interfered with my activism. Religion is often brought up in discussions, that is, "God gave us dominion over animals" and so on, but these usually lead to other threads of discussion. It's hard to debate beliefs because everyone is so unique.

Q10. FUNNY/STRANGE EXPERIENCES

Last year during a fur protest I had an unfortunate experience. Some hunters drove by and threw their freshly killed pheasant on us. It hit the person next to me and bounced off of my leg. It was a very large and beautiful bird. We got their license plate number and reported it to the police, but nothing ever came from it. This sort of ignorant backlash is just uncalled for: even "ethical" hunters would agree with me about such an incident. Other than that, my career in activism has been fairly ordinary. Not many fun or strange things happen. For all those who hope to join in the movement for the excitement . . . they had better find something else to do! It's not an adventure by any means.

Q11. LESSONS/ADVICE

The advice I would give to a person just starting into animal rights activism would be to take it slow. Many, including myself, sometimes just jump right in. It's best to be sure of what you are doing, and why you are doing it. By taking it slow, I really mean at your own pace. Don't become active because "all the cool kids are going vegan." Too many of my friends stopped eating meat just so they could fit in with a specific social group. When the friendship parts, so do their so-called ethics. Make sure you do it for yourself, above all else.

Also, don't worry about having to know all the facts. Many new activists are nervous because they feel that they just don't know as much about the issue as they should. This comes with time and experience. You can't just read a book or essay and suddenly know all there is to know. You also can't just memorize a bunch of facts. It is good to read as much as you can, though. I would recommend it. Just don't feel that you have to be an expert before getting involved. Also don't worry about being ignorant in a specific area. There are too many issues to know them all.

Q12. ADDITIONAL INFORMATION

Activists are just normal people. We are your neighbors, coworkers, and friends. Often I am frowned upon before I utter one word, because listeners know that I am a "radical activist." They automatically think that what I am going to say will be biased and untrue. I may have strong opinions, but I'm not going to twist the facts to make people see things my way.

I consider myself to be just an average American. I have views and ideas. I simply express them a little differently than many folks are used to. This is no different than the rest of history, though, as countless people have voiced their ideas or their opposition to what is going on. Activism is not new.

Deanna Krantz

Q1. BIOGRAPHICAL PROFILE

My name is Deanna Krantz. I was born and raised in Minnesota, but I spent a good part of my adult life in New York, Washington, D.C., India, and many other places on a map of the world. Helping animals has been my greatest joy and my greatest sorrow—my apotheosis and my nemesis. I cannot remember a time when I was not touched by a suffering creature: I thought this was a normal response that would be shared by all humankind. Without thinking about it (as if there was a choice to help or not to help), I simply acted to alleviate the suffering, the terror, or the fear. I now do this in India.

Q2. BECOMING INVOLVED

I do not consider myself "involved in animal rights issues." I think that the labeling of one's innate concern for animals can be sidetracked by this philosophical/intellectual pursuit. We have gone from animal welfare to animal protection to animal rights. The only progress I see is more books and more conferences: trying to redefine ourselves. I feel it is a tremendous waste of time and resources and only points to our insecurity.

Rather than donning some sturdy work clothes and working in-field where the animals need our help, we engage in endless dialogue, trying to determine if animals can even have rights. I prefer working in the jungle with my staff, performing a Caesarian on a cow in the middle of the night by firelight or flashlight, rather than meeting someone for dinner in Washington, D.C., to discuss how we can help animals. My

Deanna Krantz. Photo credit: Dr. M.W. Fox

life with animals has been a seamless journey of awe, embracing the wonder and the mystery of my fellow earth mates, but always with a call to action if anything threatened to harm them. I grew up in a relatively well-to-do, postwar suburb but had my "other life" at the farm, which was the homestead of my grandparents who emigrated from Norway. We plowed, planted, and harvested the fields. We tied a rope on Grandpa so he could make his way to the barn from the farmhouse to care for the animals and not get lost in the blizzard. Grandma would milk the cows and I would watch as her breath in the cold winter air became one with the cow's breath. She would talk to them in the soft, long vowels of her native Norwegian. No matter how hard times were, she would save some milk for the kittens and repeat over and over, "poor things, poor things . . ." (the days before spay/neuter). When Grandma broke her back and was in a cast from the chest down, she still walked to the barn to milk Red Cow, her favorite, and I would listen quietly and partake in the communion of languages I did not understand: marble-smooth Norwegian and the cow answering back. I would fall into a trancelike half sleep with the smell of hay and animals and I understood. I learned a deeper language which has held me in good stead for many years whenever doubts born of fatigue and despair crept under the door in a faraway land with so many suffering, forgotten beings. Whether it is northern snow or four-wheel drive-defying monsoon mud; whether it is a broken back on a farm in Minnesota or bone-break fever in India: you get up, put one foot in front of the other, and help one who is in need. That is joy; that is what Joseph Campbell calls "bliss."

Q3. IMPORTANT ISSUES

Since 1996 I have been working for the welfare of animals in India. It was quite by chance that I went to India with my husband, who was giving the keynote address at a veterinary conference in North India. On our journey, which began in Delhi and took me cross-country through Jaipur, Bangalore, and Cochin in South India, I witnessed animal suffering which literally brought me to my knees.

Even my tough job in New York City did not prepare me for this. As a member of the law enforcement division of the largest animal welfare organization in New York, dealing with animal cruelty cases, I thought I could face anything. I left India with a sorrow that became bolt-upright horror in the middle of the night: a neon light flashing in the darkness. A dog hit by a car, his leg hanging in ragged red by one thread of flesh. A dog on a beach in Bombay with half of his head eaten away by maggots, being stoned by men selling food in little lean-to shacks. Cows, buffaloes, horses, and donkeys mangled by lorries, left on the side of the road to die an agonizing death. They are left there either because it is illegal to kill them, or permission must be obtained from the owner (whom you can never find), or simply because they are not worth the time or the cost of medical treatment. It is cheaper to buy another animal: they are disposable.

For two hours we were unable to bypass a cattle-drive to slaughter. The ones on the road, only the sounds of their hoofs as they are death-marched to slaughter over 250 kilometers of hot, dusty roads with little or no food or water. They are tied three, four, or five together by ropes through their bleeding noses. Beaten mercilessly to keep them moving, hot chili peppers are rubbed in their eyes to get them to stand up when they collapse. Those who cannot get up are thrown into a truck—their legs broken to fit them in. Their final destination, the slaughterhouse, is Dante's hell.

Neither I nor my staff of fifteen local villagers and tribal peoples have ever thought about counting the issues that we address. We do not know from one month to the next which issues we will spend the most time on. The first issue was, of course, to have a shelter for the animals, so we built Hill View Farm Animal Refuge. Next a hospital for surgery, then a jeep/ambulance for an in-field 24 hour mobile clinic, then dealing with a massive outbreak of Foot and Mouth disease. All the while we were helping every animal that crossed our path: from the cattle stacked like cordwood in lorries to the hundreds of dogs rounded up for horrific mass killings, where they are half-strangled by tongs, and some are even plucked from their own doorsteps. The lucky ones die by a painful injection into the heart; the not so lucky die slowly while tied with wire to the bodies of dead dogs in a garbage truck. These will be sold back to their owners for five times what the killers can get from the municipality, if dead. These, and dozens more, are some of the issues we deal with regarding domestic animals.

For the last few years we have, through necessity, become increasingly involved with wildlife and forest issues because our refuge and project-site is in the Nilgiris: a "biodiversity hot-spot" with the largest remaining Asian elephant population in India. Elephants, tigers, panthers, Langur monkeys, Sambar, Malabar squirrels, Thar, Guar, and hundreds of species of birds are all being threatened through the loss of their habitats. Time is short. All of the wild adult-male tusker elephants in our area have been killed in the few years since I have lived there. My staff had names for each one. We would carefully greet them because most had become killers after years of being shot, electrocuted, having their mouths blown apart with homemade bombs put in Jackfruit, poisoned, or hacked with machetes by cash-crop owners who rape the forests. But the worst fate of all is to be captured. Such elephants live in chains, beaten into submission to carry tourists on their backs, or to promote the dubious venture called "breeding stock" for scientific research on conservation. Does torture in captivity, hidden from the public, justify our wish to conserve them? Are we doing it for their sake or ours? We have no less than fifty stories of the most barbaric, unconscionable, deliberate acts of cruelty to these wild and captive creatures, but all our efforts have met a wall of denial by powerful vested interests who do not wish to change the status quo.

Q4. GOALS & STRATEGIES

Our strategies for achieving our goals change, intensify, or modify according to the lay of the land. My project, the India Project for Animals and Nature (IPAN), is affected by the political, cultural, religious, and most profoundly by the long-standing organizational and institutional bodies which were set up to address the issues of animal cruelty and conservation. After working within these established entities to no avail, we had no choice but to expose the failure to alleviate the suffering or conserve the last of the wild. Our Web site at *www.gcci.org* (click on IPAN) details some of the issues. IPAN continues to address the situation through a holistic approach: We help the valued domestic animals, and have a project to reduce the cattle population to preserve the forests. We seek to stop cruelty to cattle during their transport to slaughter, but do not promote "saving the cows" unless there is proper sanctuary. We have witnessed and documented how lack of space, food, and veterinary care only extends the suffering of these animals. We do not preach vegetarianism because it does not address the issue of overpopulation of cattle, roaming the streets and eating plastic because they are hungry. It does not address the issue of competition with wildlife for grazing, nor the demise of the forests from overgrazing, nor disease transmission to wildlife. And it does not address the most essential issue: so long as milk and milk products are used as human food these animals will continue to suffer. The cow may take on reverential status, but concern for her fate should not take precedence over all the other animals who suffer in the streets and on the road to slaughter: goats, sheep, buffalo, burned-out bullocks, and male calves. Our Hill View Farm Animal Refuge is vegan and "no-kill," but we look for alternatives to the nonviable sound-bite solution, "save the cow." We travel to the most remote tribal settlements to spay/neuter and vaccinate dogs, and treat cattle, so the communicable diseases do not spread to the wildlife. We bring cruelty to the attention of the police and engage in enforcement of anticruelty laws in the courts. We rescue, provide veterinary services, and give permanent sanctuary to those who have been so physically or emotionally traumatized as to be deserving of a lifetime at our peaceable kingdom.

We document and report the covered-up killing and torture of wildlife and the destruction of the forests. We assist and treat many wild animals. We stop land-encroachment into the forest, the illegal cutting of trees and collection of forest products.

Q5. DEFINING SUCCESS

Our goal is to bring world attention to the plight of this rich and precious bioregion known as the Nilgiris, or Blue Mountains. It has recently been designated by UNESCO as a Global Biosphere Reserve. We fear it may be too little, too late. Research for conservation should take a backseat to an all-out, well-staffed, massive, and immediate enforcement of the laws protecting wildlife and forests, or we will only have people and cows eating their way to mono-cultural suicide. We need a consortium of international players to work hand in hand, without fear of stepping on somebody's well-heeled shoes. We need an in-field, unbiased oversight team to ensure that the programs are properly run. The only hope, the only way we will succeed, is to get beyond the cultural, religious, philosophical, and self-serving differences and simply focus on helping the animals and allowing them to live in their natural environments free of these human constraints. Impossible? No. We can stop the encroachment of people into the forests. We can encourage other means of sustainable living that does not cause pain, terror, death, or indefensibly cruel captivity.

After my years living and working with the villagers and tribals, I have found a sensitive and wise people who support this notion. They have not had a voice: IPAN has worked closely with the indigenous peoples to give them a voice. The tribals lived respectfully and sustainably for thousands of years before the area was discovered by those who wished to exploit and control the lands, the people, and Nature. We would we wise to look at our own history in this regard, in the United States, and beware lest we live to repeat it in the Nilgiris.

Q6. GROUPS & ORGANIZATIONS

IPAN's parent organization is New York–based Global Communications for Conservation, Inc. They provide our 501(c)3 charity status, handle administrative duties, and give us a Web site. All this is free of charge so that all donated monies can go to our project for the animals and the environment. Laura Utley is the founder and chairperson of this organization, and her tireless work in conservation has been an inspiration to all of us.

Q7. "THE OPPOSITION"

The "opposition" is all those people, institutions, organizations, and governmental bodies that cause deliberate harm through greed or misplaced sense of progress, without considering the soul of Nature. Nature's perfectly sensible rules do not place man in the center of the universe. Opponents are also those, who despite their good intentions, cause more harm because they do not pause before they act. I try to remember, "the road to hell is paved with good intentions." I have learned that when I listen to the trees and the tribals I am blessed with a humility that gives me wisdom and strength to fight the battle for the animals.

Q8. HEROES

I have been influenced and inspired by many writers, poets, painters, and musicians, who have themselves been inspired by the sounds, sights, and tactile wonders of animals and Nature. I cannot begin to list them all. They have no doubt sustained me.

The foundation, the grist for the mill in my early, formative years, was a gift from my family. My mother and grandmother showed me that compassion for animals was a natural thing, and, more importantly, that women could be strong if someone violated this gentle domain. When my grandfather was a little rough with a cow, my grandmother had him up against the wall with a pitchfork. When my mother saw the neighbor children torturing a bird they had tied by a string on the leg to a tree, she stood alone in protest and was slapped in the face and ostracized by the community. Her glasses went flying, but she stood fast by her convictions. My brother and I were allowed to bring in every conceivable creature we had rescued—we played hide-and-seek with a chicken running through the house. My father taught me the most important lesson, know that your compassion is not without risk and sacrifice. He taught me, through his mantralike repetition: "When the going gets tough, the tough get going." He taught me through example that we are responsible not only for our own actions but also those of others when blatant moral indifference causes harm to others.

When he allowed me to come along on a fishing and hunting trip, he was the recipient of my inconsolable anger and hurt. He never did it again; he got rid of the gun and the fishing rod. My brother was blinded when he was eighteen and when I asked him years later if he would support medical experimentation on animals if it meant he could see again, he told me that if one animal had to suffer for this, he wouldn't want to see again.

So, my advice is to follow your dream, listen to your instincts, and draw from that deep well of courage, which can make heroes of everyone who dares to take the risk. These days, many people think that heroes are only in history books. I have been fortunate to have known so many: Nigel, Mani, Madhyga, Elsi, Mary and her family, Nagaraj, Vighya, and so many others who have entered and enriched my life. Those who inspire me now are my Indian staff and friends. They are everyday proof of what I wanted to believe was possible, but dared not to believe, for fear of disappointment against the hard realities of poverty, corruption and indifference born of hopelessness. They are the most dedicated, devoted, kind, hardworking, and joyous people I have ever had the honor to work with. If there is one animal who needs our help, my staff jump in the back of the jeep, forgo food and rest, and put in a seventeen-hour day without complaint. My staff tells me that we must go up the mountain again, after two failed attempts, to rescue a severely injured old dog. It is late, we are all tired and hungry, but we succeed and now Pero lives with us and has been petted and fed and given a comfortable bed for the first time in her life. At the end of the day we feel good.

When an elephant was terribly injured during capture, my staff demanded that I become involved. We got up at five o'clock every morning to cut and collect and load sorghum, palm leaves, bananas, and cabbages into our jeep. We fed and treated him for four hours every day. We named him Loki, the messenger. We still struggle with the fact that he, and the others at the elephant camp, continue to suffer. To see the videotape, to hear Loki's howls of pain, should make one question why we bother to "save" them if it means a life of torture in captivity. Will we give up working for his liberation? No. When we feel we are losing all hope, the tribals build a fire in the donkey enclosure at our refuge and pound drums, dancing and calling out Loki's name. The young boys sing a song they wrote for him.

Q9. RELIGION/SPIRITUALITY

When confronted with what we see every day in our work, I suppose it is natural to question one's religion and spirituality. Religion is something I shy away from; live and let live. If religion is used to justify injury to others or Nature, I have no problem throwing the tea into the face of the hostess. Spirituality is a word that has been terribly overused and misused. I embrace the spirit that I see and feel in my human crusaders and nonhuman companions. I see spirit in rocks and trees and rivers. I feel it in the wind. It is part of my work and therefore cannot motivate me as something distinctively separate. Spirituality is part of it all; it does not have a category or special place in my soul. It is part of the whole that fills my soul.

Q10. FUNNY/STRANGE EXPERIENCES

Most days, humor sustains us more than religion or spirituality. It is a **M*A*S*H*** kind of humor born of a triage, foxhole experience. Sometimes it is subtle, just a look shared when we find ourselves in anatomically impossible positions trying to hold a several-ton buffalo with horns wider than most people's kitchens. Other times it is stomachs heaving and eyes filling with tears from a practical joke played on one of the staff. It relieves tension and helps to keep the balance during a particularly difficult day. There are times of almost unbearable grief when we have lost one that we love, or find that our rescue was too late, or that the suffering is too big to take in. Those are times of profound silence. We dig a grave and sit with the dogs and the donkeys, who put their heads on our shoulders and nibble our ears. The next day, life goes on and when the laughter returns it is healing. We get our greatest bonding from giggles as our new volunteers from abroad learn to adapt. During lunch, the staff uses the time to discuss various animals and plan the rest of the day. We talk about the condition of the stools of various patients, whether another treatment for lice is in order, or if we need to further treat a dog's jaw eaten away by maggots. When we look up, the new volunteer has often turned some shade of green. We laugh later. Hardly a day passes without something happening, from the mildly irritating to the dangerous, that transforms horror into humor. I announce at lunchtime that I awoke in the middle of the night with a rat in my bed, pulling out my hair for nesting material. I get no sympathy, just peals of laughter from the staff. Another time I'm walking through the dining room absorbed in thought, when I glance down and see a five-foot cobra in full attack position only a few inches from my leg. Somehow I jump straight up and flip myself over the dining room table. More laughter. Of course, one cannot live surrounded by two hundred animals and not delight in their antics. I walk into the kitchen and there is a water buffalo. I bathe with any number of creatures who are using this area as their recovery room. There is a goat recovering in the living room, one of the staff is curled up sleeping with him. The dogs have run off with the last pair of shoes. One of our cows, whom we named Plastic because we removed thirty pounds of plastic from her stomach (eaten when she was starving), comes gracefully into my tiny bedroom, gives me not the slightest recognition, and proceeds to check out my bathroom. This oddity evokes not even the slightest interest from the ten dogs on my bed! My husband walks into the office and he sees a rat taking his most important paper through the roof crack; he is shouting, and the two engage in a tug of war. The abnormal becomes the normal. The rat wins.

Q11. LESSONS/ADVICE

It is difficult to give advice because every situation has its own unique set of circumstances which is best learned by doing. Almost everyone told me that our projects in the Nilgiris would prove impossible. We were told that no one cares about animals, and we would never find staff among the locals. We met Nigel Otter; he gave us his farm for our refuge and became the deputy director of IPAN. He is the best animal handler and healer we have ever known. Hill View

Farm Animal Refuge is a model shelter for which Nigel and the staff are due full credit. We offer it as a training center for others who wish to set up similar operations elsewhere.

In this avocation, I have been a witness to my own evolution. I still marvel at how I went to India for the sake of the animals, but I stay for the love of the people I live and work with. I do not suggest that one should not be cautious, especially with some individuals and organizations that one might assume share the same vision, simply because we work in the same arena. Not all people come wrapped the same way, so beware of fancy packages. Truth and good works have a way of winning out, however endless the journey seems. There is much to be said about finding something that sustains you no matter what happens. I have 104 rescued donkeys—I love the smell of them, the sound of them, the knowledge of them just outside my bedroom window in my little renovated sheep shed. I call it home.

Q12. ADDITIONAL INFORMATION

[No response.]

Finn Lynge

Q1. BIOGRAPHICAL PROFILE

I was born in Nuuk, Greenland, in 1933. My father was a native Greenlander, and my mother was Danish. My two sisters and I spent our childhood, including the World War II years, in Greenland. I spent my secondary school years in Denmark, followed by university studies.

I converted from the Lutheran to the Catholic version of Christianity at the age of nineteen, and five years later entered the training for priesthood. My training included one year (novitiate) in France, three years (philosophy) in Rome, and four years (theology) in the United States. I spent thirteen years in active ministry in three countries (the United States, Denmark, and Greenland), after which I received a dispensation and was married.

My career shifted then to social work, radio broadcasting (administration), and politics. I was elected to the European Parliament for Greenland, as a consultant to the Danish Foreign Ministry in Copenhagen. Later I became a senior advisor to the Greenland government.

Q2. BECOMING INVOLVED

The animal rights issue was placed on my table for the first time when I had entered the European Parliament in Strasbourg, France, in 1979. I grew up in a community where sealing and whaling were the bread and butter of daily life. But here I suddenly found myself in an assembly where hunters were treated more or less as murderers. I wouldn't say it came as a complete surprise. In Greenland we had

Finn Lynge.

heard about the antiwhaling and antisealing campaigns for some years. All the same, hearing about people with such "strange and unnatural attitudes" (as it appeared from an Inuit perspective) was one thing, but facing them directly was something else. Since then, the animal rights issue has stuck with me for good.

Q3. IMPORTANT ISSUES

What really interests me is the sociocultural roots of the issue, or to put it in another way: In a given context, what exactly triggers people's reactions? Among the animal rights advocates, I have always been struck by their general lack of knowledge of other people's living conditions, and, most of the time, their lack of interest in the issue of cultural relativism. This issue seems to attract absolutist minds. That fact in itself, I find puzzling. Why is it that way?

Q4. GOALS & STRATEGIES

Strategies for achieving a goal? Well, inasmuch as my goal is to foster a better cross-cultural understanding, the best thing to do is to support existing positive trends in the field. In that connection, much good can be said about the World Conservation Strategy of 1991 "Caring for the Earth," the Rio Declaration and Agenda 21, the ongoing work in connection with the Biodiversity Convention, and so on. Much sound thinking is being tabled.

The question is to what extent it really sinks in when it comes to the public—the vast groups of television-watching and tabloid-reading public, people who are not apt to try and think along lines other than the well known.

Personally, over the years, I have been engaged in public speaking and taking part in debates.

Q5. DEFINING SUCCESS

[No response.]

Q6. GROUPS & ORGANIZATIONS

After spending five years in the European Parliament, I became involved with Arctic hunters' NGOs (nongovernmental organizations) such as the Inuit Circumpolar Conference (ICC), a United Nations-accredited NGO for the 110,000 Inuit of Greenland; Arctic Canada, Alaska and Chukotka (Far Eastern Russia); and Indigenous Survival International (ISI), an organization for both Inuit and Indian hunters and trappers of Northern Canada and Alaska. For those two organizations especially, facing up to the animal rights activists of North America has always been a priority, defending the UN and Rio-established right of indigenous peoples to pursue subsistence hunting and to market products therefrom. Later, I worked with the Copenhagen-based International Working Group for Indigenous Affairs (IWGIA); the Danish section of the World Wide Fund for Nature (WWF); the Brussels-based European Bureau for Conservation and Development (EBCD); and entered the International Union for the Conservation of Nature and Natural Resources (IUCN), joining the work in both the Inter-Commission Task Force on Indigenous Peoples and the Law Commission Ethics Working Group. I helped to found a Danish NGO by the name of Nature and Peoples of the North, which in turn has helped to build up the Russian Association for the Indigenous Peoples of the North (RAIPON), headquartered in Moscow.

Once employed by the Danish Foreign Ministry, I was assigned a seat in the delegation to the International Whaling Commission (IWC), an organization which may be termed the most important international forum for animal rights on a high political level. I served in the Danish delegation to the IWC from 1987 to 1997.

Q7. "THE OPPOSITION"

I feel that I have two camps of opposition:

First, those who don't understand, or don't want to understand, the legitimacy of the hunting cultures of old. They do not care about the needs of contemporary hunter-gatherers. There are many hunting cultures around the world, millions of people, most of whom are poor and politically oppressed. The International Fund for Animal Welfare

(IFAW) never lends those people a helping hand. Neither does The People for the Ethical Treatment of Animals (PETA). Greenpeace is a fence sitter in this matter. In these three organizations, their love for animals prevents them from loving people. I have no use for such organizations. I am prepared for a public showdown with them any time.

Second, the other "opposition" is that group of hunters (hopefully rare!) who don't care about biological sustainability of the hunted stock, or about the basic animal welfare measures to be taken in any hunt. Both hunters and antihunters have their share of narrow-mindedness and insensitivity, as indeed does every group of humans.

Q8. HEROES

Who are my heroes? Gro Harlem Brundtland, Teilhard de Chardin, Chief Seattle, Job, Fyodor Dostoyevski, Nelson Mandela, and Thomas Berry. Thorough thinking, doing your thing, facing the consequences, that's what makes or unmakes a hero.

Q9. RELIGION/SPIRITUALITY

Yes, religion and spirituality is an essential part of my life. I couldn't function without it. I am a Roman Catholic, which doesn't mean that I am enamored with the Roman Curia. Much stifling authoritarianism is found in church traditions of all times. I, for one, am happy that the Spirit blows wherever it wants. The Good Lord has a very ecumenical mind, I am sure.

Speaking of the "Good Lord" . . . in the field of animal rights, I keep reminding myself about what is written on the first page of the old Book: "[A]nd behold, everything he made was good." It was very good indeed. This includes the lion that is about to break the zebra's back, the killer whale tearing up the defenseless humpback, and the cat going for the mouse. Predation is part of creation. Predation is good, very good indeed. Humans are also predators.

Q10. FUNNY/STRANGE EXPERIENCES

I find it striking that the one people on the surface of the Earth who have absolutely no option other than to live as predators—the Inuit of the Arctic—are also the one people who have never engaged in wars, who have no army, and who do not engage in military training. Spilling the blood of a human? No! Killing an animal for food and clothing? Certainly. We are made that way, and we love and respect everything

He gave us. Especially the seals and the whales. They are great creatures to watch.

And they taste so good.

Q11. LESSONS/ADVICE

Learn to enjoy being contradicted and to contradict in return. This is not for the purpose of digging ditches as canyons already exist. We have this great debate for the purpose of building bridges.

Don't be afraid of being politically incorrect if you are convinced you're on the right track. Don't attach too much importance to what other people think. Keep on learning, all life long, about other nations and other cultures, their thinking and their ways. Cultivate the virtue of both giving and taking. Be open. Be positive.

Q12. ADDITIONAL INFORMATION

[No response.]

Kathleen Marquardt

Q1. BIOGRAPHICAL PROFILE

I am Kathleen Marquardt, born in Kalispell, Montana. I enjoy scuba diving, skiing, fiber arts, and reading.

Q2. BECOMING INVOLVED

I became involved in animal rights issues when my younger daughter, in junior high, came home from school (in a suburb of Washington, D.C.) one day. Her Science class had an animal rights guest speaker, who had asked the children how many of them had fathers who hunted. No one raised a hand (not unexpected in such an urban area) except for my daughter, who said that her mother hunted. The speaker then told my daughter that her mother was a murderer. My daughter came home in tears, not wanting to ever go back to school again (who would, if her mother was the equivalent of Jeffrey Dahmer!).

I was livid. This woman had pushed her radical values onto vulnerable children. Who was she and what was PETA (People for the Ethical Treatment of Animals), her organization? Seeking the answers was what got me into the fight. Before that I was not a joiner, I had belonged to no organizations. My philosophy of life was to live and let live.

Q3. IMPORTANT ISSUES

The two issues that I work on the most are the strong-arm tactics of animal rights groups and the religious basis of animal rights.

First, the strong-arm tactics: why do animal rightists even need to use coercion? If their cause is so wonderful, so right, and so pure, why would they need to resort to terrorism, intimidation, and brainwashing? Wouldn't we all be jumping willingly on to their bandwagon? Don't we want to be on the side of justice? Animal rightists have promoted their beliefs and cause since the late 1960s. They have preached their doctrines across America. And what size is the flock produced by such proselytizing? The usual 3 percent to 4 percent, the fringe (some would say the lunatic fringe).

Because their use of legal, ethical, and proper means was only able to drum up this small following, they have begun to resort to illegal, unethical, and downright abhorrent means to cram their beliefs down our throats.

Using brainwashing, bullying, and terrorism, they now hope to bring us into the fold. Animal rightists go into our schools ostensibly to talk about humane care of animals. Instead, they feed our young and impressionable children fuzzy, feel-good lies and distortions. "Meat is murder"; "Animals are our friends, we don't eat our friends"; and so on. These sound bites of unreality are fed to our children whom we send to school to listen and to learn. Parents should be scared out of their complacency over what their children are learning in the classrooms, if they think the children are learning reading, writing, and mathematics.

Animal rightists warn fur coat wearers: If you walk out of your door in a fur, you should expect to be heckled, spray-painted, spat upon, or worse. What group of people wears fur coats? Mostly older women.

Why aren't these animal rightists using the same tactics on groups that wear leather? Leather is no different than fur—it is the skin of a dead animal. But no animal rightist is going to be stupid enough to voice his or her displeasure in a biker bar, even though the aggregate of dead animal skins in that bar is probably far greater than at a New York tea party. And why won't they confront bikers? Because fur coat wearers are usually small, nonaggressive women, whereas bikers are large, willing-to-fight men. Animal rightists are bullies; they use their intimidating tactics on those who are least willing to fight back.

Last but not least, the strong-arm tactic is most blatant in the bombings of research labs and threats to researchers' lives. Animal rightists want the world to believe that they are kind and caring beings, yet they are willing to destroy laboratories, to risk the lives of researchers and laboratory animals in order to push their beliefs (religious beliefs) down our throats.

Yes, religious beliefs. In order to believe in animal rights, one must suspend critical thinking. Animal rightists believe with a zealot's fervor that animals are equal to or better than humans, and, in fact, that it would be a good thing if man became extinct and left the earth to the animals. "Mankind is the biggest blight on the face of the earth," says Ingrid Newkirk.

"If you haven't given human extinction much thought before, the idea of a world with no people in it may seem strange. But if you give it a chance, I think you might agree that the extinction of Homo sapiens would mean survival for millions, if not billions, of Earth-dwelling species . . . Phasing out the human race will solve every problem on earth, social and environmental," writes Les U. Knight (pseudonym) of the Voluntary Extinction Movement.

Why should others care about the doings of animal rightists? I think it should be obvious from the previous paragraphs. The general public should care about these strong-arm tactics because coercion is always wrong. America is a free country. We have First Amendment rights; there is no need to use force. But the animal rights people have gotten all of their adherents through proselytizing.

Now they believe that they must resort to threatening little old ladies wearing fur coats, bombing research labs, and threatening researchers and their families. They haven't realized that, just as one can't legislate morality, you can't force religion down people's throats (remember the Spanish Inquisition?).

Animal rights is a religion to many people.

The premise behind the animal rights movement is that animals are equal to, or better than, humans. What is this based upon? Scientific fact? Not hardly. In what way are animals better than humans? Rightists will say that it is because humans can be cruel and animals can't, and therefore animals are better species.

Actually, animals can be cruel—watch a bear play with a salmon, or a cat with a mouse. But that is not the crux of the issue here. The moral beingness of humans is.

As of this moment, humans are the only moral beings on this planet. We are the only species that comprehends right and wrong, and can act on those concepts. Animals are amoral: not knowing right from wrong, acting out of instinct or by training. The only species that can be held responsible for its actions is Homo sapiens.

Our moral cognition is what animal rightists hold against us. The religion stemming from the animal rights movement is one of negativity—

hatred of the human species—but is manifested in reverence for animals.

To believe in something that cannot be proven requires suspension of critical thinking (a determinant of a religion). Animal rights' preeminent U.S. philosopher Tom Regan even admits as much in *The Case for Animal Rights*, when he discusses how to deal with the "rights" issue. "I am not even certain that they can be settled in a rationally coherent way." But rationality is not mandatory to animal rights theory for Regan goes on to say, "The rights view, I believe, is rationally the most satisfactory moral theory. Of course, if it were possible to show that only human beings were included in its scope, then a person like myself, who believes in animal rights, would be obliged to look elsewhere" (Tom Regan, "The Case for Animal Rights," in Peter Singer, ed., *In Defense of Animals* [New York: Harper and Row, 1986]).

This brings one to the conclusion that animal rights must be a religion, because there are no other grounds on which to support the belief that animals are equal to or better than humans.

Q4. GOALS & STRATEGIES

In the short term, I seek to expose these strong-arm tactics to the general public. Americans don't like such tactics; they will reject these attempts out of hand. I also seek to keep their religion from being promoted in our schools (just as we keep other religions out).

In the long term, I want to educate the majority of the populace on what the animal rights issues are all about.

Q5. DEFINING SUCCESS

I will have succeeded when the general public is divided into two groups on this issue: those who acknowledge that animal rights is their religion of choice at the present time, and those who understand and reject the animal rights belief system.

Q6. GROUPS & ORGANIZATIONS

[No response.]

Q7. "THE OPPOSITION"

Obviously the opposition is the animal rightists who use coercive tactics or who lie to get their message across (especially to children), for example, "meat is murder." According to *Webster's Dictionary*, murder is the unlawful killing of another *human being*. So, unless we are eating other human beings, this sound bite is another lie.

I am also against those who fight for rights for animals. "Rights" is a concept conceived by moral beings for moral beings: those who can understand the difference between right and wrong and can choose to act on those concepts. Animals are not moral beings; they do not understand right and wrong, and cannot be held responsible for their actions. Therefore they do not have rights. We human beings have a responsibility to treat animals humanely.

Q8. HEROES

Dr. Dixie Lee Ray is one of my heroines. She was an oceanographer who believed in using sound science and reason to conserve species and their habitat. Yes, she was willing to be politically incorrect in her dealings with environmental issues in order to support scientifically based means for dealing with environmental issues—to use sound science first to determine if and what the problems are and secondly to correct them.

Q9. RELIGION/SPIRITUALITY

Religion to me is a very personal thing (or should be). We all have our beliefs, whether they are organized religion, science, or one of the New Age philosophies that are not yet formalized religions.

It is the basis of the animal rights movement: a belief that animals are equal to or better than humans and that humans are the scourge of the earth.

Q10. FUNNY/STRANGE EXPERIENCES

When we were living in Washington, D.C., and went grocery shopping, if an animal rights activist was in the store, he or she would follow us around calling us filthy names and threatening. When we were in Montana, I was approached in a video store by a man who claimed to be with the group Earth First!, spiking trees. He told me he could do the same to me. And these people call themselves kind, gentle, caring individuals—obviously this is true only if it concerns animals or the people who agree with them.

Q11. LESSONS/ADVICE

Understand that there are two kinds of thinking: emotion-driven or reason-driven. Animal rights activists are emotion-driven thinkers, but not all emotion-driven thinkers are animal rights activists. You can appeal to those who are reason-driven: Give them the facts, and let them figure things out for themselves. You can appeal to many emotion-driven folks who don't believe in the animal rights religion. But you must accept that you can't get everyone who hears your facts to come to what you consider to be the correct conclusions. Just accept that there will be a small percentage of people—the fringe—who reject critical thinking totally. But you can also know that a majority of people will use common sense and civility, and thus be on your side.

Q12. ADDITIONAL INFORMATION

[No response.]

Pat Miller

Q1. BIOGRAPHICAL PROFILE

My name is Pat Miller. I am the owner of Peaceable Paws (training, writing, and consulting) and trainer for Peaceable Paws Dog and Puppy Training. We offer various types of training, but our primary focus is family dog training. We use only gentle, dog- and owner-friendly training methods based on the modern, scientific principles of operant conditioning.

Prior to starting my own business five years ago, I worked for the Marin Humane Society for twenty years: eighteen years as a humane officer, and the last ten also as director of operations.

I am also a writer. I write two or three articles a month for the *Whole Dog Journal*, and I am a regular writer for, among others, the *Whole Cat Journal*, the *Whole Horse Journal*, and Tuft's University's *Your Dog and Catnip*. I also have a book, *Positive Dog Training* (2001).

I was born in 1951 in Springfield, Illinois. I enjoy spending time with my husband Paul, and our four-footed kids: four dogs (Josie, Tucker, Dusty, and Katie), two cats (Jackson and Gewürztraminer), and our horse, Rafiki. I also enjoy reading, bird-watching, photography, hiking, and gardening (when I have time . . . !).

Q2. BECOMING INVOLVED

I have loved animals ever since I was a small child. I always wanted to be a veterinarian (until teenage rebellion and Organic Chemistry put my college career on hold until I matured and returned to school, first to get my Associate degree in Administration of Justice in the early

Pat Miller.

1980s, then to get my Bachelor's degree in Business Administration in 1996). My favorite book as a child was *Beautiful Joe* (in which the human heroine, Laura, is an animal activist and the animal hero is an abused dog), followed closely by *Black Beauty*. My career in animal protection exposed me daily to abandoned, abused, and neglected animals, which fueled my sense of outrage over the injustices we as a species inflict upon them. Reading Peter Singer's landmark book, *Animal Liberation*, helped me form my outrage into cohesive philosophies. And meeting my husband, Paul (who also works in animal protection), in 1982, convinced me to make animal protection a lifestyle and my life's work.

Q3. IMPORTANT ISSUES

The two issues that are most compelling for me are pet overpopulation and the abuse suffered by farm animals. Pet overpopulation is the root cause of most companion animal abuse (I include our equine friends under the definition of companion animal). Because they are so easy to obtain, their lives are not valued. They are abused and neglected at will, and discarded on a whim. Farm animals suffer legalized abuse by the millions on a daily basis, from the unspeakably inhumane practices of factory farming to the horrors of livestock transportation and slaughter. Other people should care about this because if we turn our backs on the suffering of animals we become calloused to the pain of all living things. If we are to be a humane society, we need to cultivate a pervasive respect for all life.

Q4. GOALS & STRATEGIES

My dog training methods are designed to teach owners how to communicate with their dogs without using pain, intimidation, or fear in order to foster a relationship of mutual trust and respect between dog and human. I support and proselytize for spay/neuter and responsible companion animal caretaking. The twin approaches of spay/neuter and training are the two most important keys to addressing the challenge of pet overpopulation. Spay/neuter, obviously, is necessary to reduce the numbers of homeless and unwanted animals. When they are scarcer, they will be held in higher value. Training is also an important key, because as shelters succeed in reducing the numbers of surplus puppies and kittens at shelters, they are faced with a population of adolescent and adult animals who are surrendered by their owners (or unclaimed strays) because the social contract between companion animal and human has been broken. According to recent research, the most common reason for this breach, at least with dogs, is a lack of training. A study conducted by the National Council on Pet Population, published in the *Journal of Applied Animal Welfare Science* in 1998, found that a staggering 96 percent of the dogs surrendered to shelters by their owners had not received any obedience training.

I am an active and vocal member of the Association of Pet Dog Trainers, a national organization that promotes the use of positive dog training methods. My writings actively support the "respect for life" philosophy that I embrace. I was also a founding member of CHAIN (The Collective Humane Action and Information Network), a non-profit organization that promotes professionalism in the field of animal law enforcement, and for eleven years I was editor and publisher of CHAIN's quarterly magazine, the *CHAIN Letter*. I don't hesitate to speak out on animal issues on the Internet, in articles, and in any other forum that seems appropriate.

My advocacy for farm animal issues is expressed through my consultant work with the Humane Farming Association, a national non-profit organization that works to improve the quality of life for farm animals and to educate people about farm animal issues.

Q5. DEFINING SUCCESS

On both issues, my ultimate goal is to see every individual animal respected and valued as a sentient creature with an inalienable right to a quality life.

With pet overpopulation, this means reducing the birthrate of companion animals such that they are in demand, valued, and cared for rather than treated as disposable and tossed away as garbage. We can say we have succeeded when animals are no longer neglected in backyards, abused, abandoned, and euthanized by the millions at animal shelters simply because there are too many. This would include committing resources toward fixing the physically and psychologically broken ones rather than dumping them in a category labeled "unadoptable" and pretending that their deaths don't matter, as is all too common in the so-called "No-Kill" movement that has gained a strong foothold in this country.

The logical ultimate goal for farm animals would be to get people to stop eating them. Short of that (because that is not likely to happen), it would be at least to ensure that they are raised in natural environments that meet their physical and psychological needs for appropriate food, water, shelter, exercise, and companionship. And if they must be killed, it should be done in a truly humane manner, without stress or suffering. This would mean a return to family farms, where animals are killed without the stress of transport, livestock sales, and assembly-line slaughter methods.

Q6. GROUPS & ORGANIZATIONS

I work with the Santa Cruz SPCA (where I was a board member until my 1999 move to Tennessee), the Humane Farming Association and the Association of Pet Dog Trainers. I am also a founding member of the Tennessee Association of Positive Pet Trainers. I was a member (and one-time president) of the California Animal Control Directors' Association and a member of the Society of Animal Welfare Administrators until I left California. I have done pro bono training for the City of Chattanooga's new Animal Services Division, and am volunteering my services with the Chattanooga Humane Educational Society.

As I mentioned, I was a founding member of CHAIN and longtime editor/publisher of the *CHAIN Letter*. I also worked with the local group LAW (Licenses for Animal Welfare) to promote Chattanooga's first Spay Day USA event in the year 2000.

I am still a member of the Marin Humane Society, where a part of my heart will always reside.

Q7. "THE OPPOSITION"

I prefer to think of them in terms of the "unenlightened" rather than the "opposition." Obvious groups are the livestock industry, and any other group that commercializes and exploits animal life—horse racing, pet stores that sell live animals, breeders who raise them for commercial purposes, importers who bring exotic animals here from other countries for commercial sale, dogfighters, cockfighters.

Q8. HEROES

My husband, Paul Miller.

Diane Allevato, executive director of the Marin Humane Society.

Phyllis Wright, previously vice president for Companion Animals of the Humane Society of the United States (now deceased).

My mother.

Albert Schweitzer.

St. Francis of Assisi.

Q9. RELIGION/SPIRITUALITY

I am not a follower of any organized religion. I do believe that I am a spiritual person and that this motivates my work. I believe that we are all connected through some sort of universal consciousness, and that it is this connection that compels us to care for and about other living beings.

Q10. FUNNY/STRANGE EXPERIENCES

At one point in time, my husband and I were doing an in-depth study of animal sacrifice, including Satanism and Santeria. We had the opportunity to meet Anton LaVey (now deceased), then head of the Church of Satan. A mutual friend was doing the introduction—we were going to visit LaVey at his house in San Francisco. Paul and I arrived first. It was nighttime, and as we leaned against the lamppost in front of LaVey's menacingly purple Victorian home, we realized that it was a full moon–a very meaningful time of the month for Satanists. Our friend finally arrived before we got too spooked, and we spent a very entertaining evening with LaVey, chatting over his gravestone-cum-coffee table, and condemning animal abuse. LaVey insisted that he did not believe in animal sacrifice and that he was a confirmed animal lover.

Q11. LESSONS/ADVICE

It is easy, as you learn more and more about the abuses that animals suffer at human hands, to become overwhelmed by a sense of futility. If you are going to help animals, you have to be able to find and keep a sense of perspective and balance. If you lose that, you can do the animals no good, and you certainly do yourself no good. Find ways to stay centered. Develop strong support systems—people who share your beliefs, and activities that reconfirm your connection to all that is good in the universe. Remember to take time for yourself and the human and nonhuman animals you love—if you let yourself become too immersed in the struggle you can lose both those loved ones and yourself.

Q12. ADDITIONAL INFORMATION

[No response.]

Laura Moretti

Q1. BIOGRAPHICAL PROFILE

I'm Laura Moretti. I was born in Massachusetts, on a U.S. Air Force Base nearly forty-three years ago; my father retired as a colonel. I was raised across the globe from Europe to South America. These childhood experiences in other countries not only triggered my animal rights activism, but also helped me to enjoy the differences between animals and humans, as well as between other cultures and our own.

I enjoy history, politics, Native American philosophy, writing fiction, and spending time with the two cats (who live with me) and a horse that I rescued from slaughter about seven years ago. I also enjoy close family ties, good books, and really good movies. Although I can't truly say that I enjoy animal rights activism, I'm thankful that it is in my life. Ignorance is not bliss.

Q2. BECOMING INVOLVED

I have very vivid memories of certain shocking events that took place in both Spain and Bolivia when I was six and nine years old, respectively.

When I was six, a large white husky-type dog followed me home from school. It seems that I was always bringing home animals, from beetles to frogs and gophers to snakes, plus cats and dogs. So I didn't think anything of this huge white dog who trailed me along the pathway. He stood nearly at my shoulder and was covered in a thick, dense coat. In my view, he was utterly massive and strikingly beautiful.

But when I got home and Mother opened the door to allow me entry, she seemed suspiciously wary of my companion. She ordered me in quickly and closed the door, shutting the dog in the open yard behind the screen. I don't recall being told there was any reason for her caution, but I vividly remember watching from a window as the dog hiked to the top of a nearby plateau. I was curious about why I should be so wary of him as he had followed me all the way home without incident.

On the road, not too far away, a patrol car came into view. And I watched, mortified and helplessly, while the dog was gunned down in the street. Later I would learn there had been a rabies scare, and all stray dogs were to be shot on sight.

In Bolivia, I watched again with that morbid dread that came from sheer helplessness and inexplicable outrage, while a dog was strangled to death in the street, to become a family's evening meal. I saw llamas viciously beaten into carrying heavy loads. I glimpsed the old woman across the street, the one who lived in the adobe hut with her children, cows, pigs, and chickens, wring the necks of small, screaming rabbits, whom she later sold at the open market.

Dead dogs and livestock littered the riverbed that I had to cross every morning on my way to school. Soon I found a private place on the side of the mountain behind our house where I'd take their remains, bury them, and pray for their souls despite the local Catholic Church's claim that animals didn't have any souls.

I can't say it was any single event that put me on this road of lifelong activism, but I knew at a very early age that there was something terribly wrong with the way humans viewed and treated animals. I empathized with their helplessness and innocence. So it was only inevitable, I guess, that when we returned to the States and I watched a news documentary showing the brutal slaughter of seal pups in Canada, that my life would change in that moment and never be the same.

Q3. IMPORTANT ISSUES

Perhaps like a lot of animal defenders, it was years before I made the connection between the atrocities committed against, say, seal pups in the name of fashion, and veal calves in the name of food. I picked up a book by Ruth Harrison entitled *Animal Machines*, and was so mortified by what I learned that I became a vegetarian that very instant. How I had missed the direct link between the pig and the pork chop is something I am only recently beginning to fully understand and appreciate.

I think that the plight of "food animals" is probably the single most important issue of our times. Every year, nearly 10 billion animals are butchered for human consumption in the United States alone. By sheer numbers, the treatment and exploitation of animals in the name of food should demonstrate that we must end this practice. This flesh-eating industry destroys the environment, human health, and animals' lives in unfathomable numbers.

More than that, ending our consumption of animals (unlike ending war) is one thing that each and every one of us can do to improve our health, protect the environment, and abolish the abuse, suffering, and exploitation of animals. I truly believe that a higher plane of consciousness rises in the individual who leaves off eating animals, an awareness that opens the mind and empowers the soul, which allows us to embrace the true essence of humanity: our empathy.

Q4. GOALS & STRATEGIES

A single image changed my life: a seal hunter kicking a seal pup in the face until it was dead left me with a sense of purpose from which I would never escape (nor would I want to). A single picture of a crated calf in a veal barn altered forever my behavior. I still believe in images, and so I spend my time and my life finding ways to bring those images to the widest audience possible.

In my youth, I did that by designing and posting my own flyers, purchasing newspaper ads, writing letters to the editor of local newspapers, selling magazine articles, and publishing my own books, such as *All Heaven in a Rage: Essays on the Eating of Animals*. I also founded and edited an international animal rights magazine, *The Animals' Voice*, to bring awareness to the plight of animals worldwide. Now, my outreach is encompassed in the Web site, Animals Voice Online, with three sister sites in development. It is, I believe, the most cost-efficient, far-reaching way to educate others about the plight of animals.

The ironic thing is that I wanted more than anything to work in the local school system and teach the psychology of prejudice, of humans and animals. In college, I was told animal rights was a bogus concept, and I dropped out. In a roundabout way, you could say I found another path to my initial dream. With a computer keyboard, I have become a teacher.

Q5. DEFINING SUCCESS

On a certain level, I have to say that we have already succeeded. I've been active in the defense of animals for nearly three decades, and I have seen a lot of changes. Granted, we have a *long* way to go. But, as some philosophers have noted, there are no longer quotation marks around animal rights in the mass media. Success appears in such little things as the disclaimers at the ends of movies, the advertisements of fake furs, or the little red hearts on the dinner menus that denote vegetarian meals. There was a time when I was eyed with the same kind of wariness given the big white husky-type dog of my youth. Success, to me, has come by way of a movement strong enough to stand on its own till the job of animal liberation is done. That movement is well on its way, and there seems to be no stopping it.

Q6. GROUPS & ORGANIZATIONS

Over the years, I've been affiliated with or worked for a number of organizations: The Fund for Animals, the Compassion for Animals Foundation, Last Chance for Animals, In Defense of Animals, Farm Sanctuary, The Animals' Agenda, and, of course, *The Animals Voice*. With the exception of my undercover work with the New Jersey State SPCA, my involvement continues to remain steadfast: merging words with images in ways that, I hope, cause people to think and change.

Q7. "THE OPPOSITION"

One could make a serious argument that our adversaries are the industries that exploit animals for profit: agriculture, research, entertainment, and so on. But our real opposition is the consumer who demands of these massively financed monsters the products they sell.

Nowhere is that clearer to me than in the food industry. There's a little known place near Vernon, in California, where pigs are slaughtered on a daily basis. Across the street from Farmer John's Meat Packing Plant is a McDonald's. If you listen closely, you can hear the pigs screaming from there. You can smell their blood. But I've sat on the curb, with pigs screaming in the background, and watched the passers-by. They only seem impressed by the plant's façade: an imposing wall painted with pink pigs frolicking in green pastures.

There is a reason there are hidden cameras at such plants. There is a reason why the confines of such a place must be disguised from public view. There is a reason you can't fight your way into a slaughterhouse, though none to stop you from taking a tour of a cornfield or an apple orchard, and that's because even the giants that derive their profits off the backs of animals know that knowledge is power. What people know will hurt them.

And so it is the consumer who must be made aware of the terrible sound and suffering that goes on behind closed doors and deceivingly painted walls. They're the little people, as the saying goes, who make the big people big.

Q8. HEROES

The people who inspire me the most are, ironically, the people I know the least. They're the trench workers, those unsung heroes whose lives have changed because of some image or experience, and they chip away at the machine that has ground up not just animals, but our human essence in the process. They're the lone activists who rescue animals, who wage nearly silent wars in their corners of the globe, who feel alone and yet find the strength, through animals' eyes, to put in one more minute, one more hour, one more day, on behalf of the living.

One of my named heroes, however, is John Walsh, international projects director of the World Society for the Protection of Animals. In his lifetime, he's been personally responsible for the rescue of tens of thousands of victimized animals: from natural disasters to human wars. Very few organizational leaders are also workers in the trenches. He's one. He could sit in a comfy office chair and dish out orders, if he wanted to. Instead, he travels to places like Kuwait, and with Scud missiles firing overhead, risks his life to feed the trapped animals at the zoo. My respect for him is immeasurable.

Q9. RELIGION/SPIRITUALITY

I left the church and God when I was living in La Paz, Bolivia, by wondering what sort of creator could visit upon helpless, innocent animals the kinds of atrocities I had witnessed as a child.

In recent years, I've come to realize that I replaced that religion, not with atheism, as I once believed, but with pantheism. The belief that there's a spirit, a soul, in all there is: mountains, rivers, the way geese migrate. It is this newly found religion, if you will, that now inspires me and secures my dream that one day we shall achieve, as Einstein puts it, the goal of all evolution: nonviolence, the highest ethics. To achieve such a goal we must, I am certain, include animals in that moral sphere.

In the Christian Bible, God created the Garden of Eden. In it he put vegetarian humans. My purpose in life, through my newly discovered "religion," the sacredness of life, is to bring us back to the Garden's Gate.

Q10. FUNNY/STRANGE EXPERIENCES

Can't think of any. Moving, yes, but not funny or strange.

Q11. LESSONS/ADVICE

Quite simply, I'd tell newly born activists to follow their hearts, and to avoid the one thing that can trouble it: the politics of activism and the ignorance of the masses. Our personal goals as activists are to rise above whatever differences we may have with one another, and to remember our roots: Most of us weren't born animal rights activists. The lot of us have worn leather shoes, eaten animal flesh, visited a zoo.

A friend asked me recently what I thought about the fact that he wasn't a vegetarian. I thought about it, but not for long. "I know how you feel," I finally answered him. "I used to eat calves liver and onions." I differed from him only in the depth of my awareness. The trick, I realized then, was to find ways to open his mind the way mine had been opened without shutting it first.

The best way to do that, I think, is to remember where you've come from, to remind one's self of what it is that we have in common with our fellow humans, to "forgive what they know not," as we've forgiven ourselves. Then to be the voice of animals, in ways that our fellow humans can hear them, as we ourselves once heard them and will continue to hear them for the rest of our days.

Q12. ADDITIONAL INFORMATION

Our roots are not just personal. Human history goes back a long way. As an insatiable student of history, I take note of the path we're on and how far we've come. Yes, granted, we have more ground to cover than we've crossed, but there was a time, looking back, when there was no break on the road into which human beings defending animals had taken.

It can and I speak from experience be a most demoralizing and painful journey, but as I once wrote in an essay titled, "Home Is a Wounded Heart":

> Shut our eyes? We cannot. Wish that we could? We would not. We know healing and change begin with the truth. Sadly, the truth begins with a broken heart. But better to have a wounded heart, a sick and hurting heart, than to have none at all, to feel for nothing, to care for no one, to live for death. Keep fighting the good fight.

Ingrid Newkirk

Q1. BIOGRAPHICAL PROFILE

I was born in Kingston-Upon-Thames in Surrey, England, in 1949. I enjoy my coworkers' passion about their jobs, cryptic crossword puzzles, comedy of all kinds, the presence of my human companion, mathematical puzzles . . . anything that takes my mind off of cruelty to animals.

Q2. BECOMING INVOLVED

I was drawn to animals from childhood, even as one might be drawn to Legos or art. It was a natural and commanding part of my life.

In adulthood, it was bad luck that led me to animal work. I was enjoying idiotic pleasures when a neighbor moved away, abandoning nineteen cats. As a result, I visited an animal shelter, and found the conditions to be very poor. From then on, I have moved through various spheres of animal suffering, and my own evolution of understanding.

I became a vegetarian after finding a starving pig in a barn, and later that night, finding that I had pork chops in my refrigerator. I put two and two together. Like most people, I can be a slow learner, and it took time to realize the extent of my participation in animal pain and death.

Q3. IMPORTANT ISSUES

I would rather be pulling injured animals out of gutters, but I concentrate on administration: keeping the whole "machine" functional

Ingrid Newkirk. Photo credit: PETA

through the ups and downs, personalities, unexpected developments, and financial/legal aspects. I care because I want the organization to thrive, because if it does thrive, more people will learn how they can help change the animals' world for the better . . . sooner.

Q4. GOALS & STRATEGIES

My strategy is to make sure that we do not miss opportunities and that we have done our homework (research, reading, investigation). I write letters, editorials, articles, and books. I oversee the organization of protests, demonstrations, and other events. I make suggestions (and sometimes insist on certain changes), to keep the tone consistent with our goals. I try to discourage pettiness and in-fighting. And, of course, I make sure that we pay the bills and share our ideas.

Q5. DEFINING SUCCESS

The ultimate goal can never be achieved. It is an end to all cruelty. Like the pursuit of world peace among human beings, it is unachievable; so there is no retirement and no end. It will never be over.

Q6. GROUPS & ORGANIZATIONS

I work with People for the Ethical Treatment of Animals (PETA). We will work with almost any group except for the Nazi party and the Ku Klux Klan, wherever our paths cross on any issue. All that matters, for the most part, is what we share in common on the issue, not where our paths diverge. We have worked with hunters to pass an antitrapping law, for example; and we form ad hoc coalitions with environmental, animal rights, and human rights groups. We have helped the right-to-life groups fight the National Organization for Women over the right to protest. We believe that no two individuals (let alone groups!) can agree on everything, or even on their main goals, so it is important for us to agree on what we can.

Q7. "THE OPPOSITION"

Anyone who vigorously defends animal exploitation and fights positive change for animals (for example, furriers, butchers, animal experimentation trade groups) would be a current member of the opposition. However, almost everyone has the potential to change. Years ago, I met a man who used to make fun of me for refusing to eat steak and eggs with him. He supported the National Rifle Association and considered vegetarianism an eccentricity. He liked dogs, but that was about it. Over the years, he changed when I wasn't looking. Today, he has left the NRA and is very critical of its position on hunting, and he is a vegan and an outspoken critic of fishing. Devout racists have become advocates of racial harmony, warmongers have become peaceniks, people who experimented on animals—like Donald Barnes, who irradiated monkeys for the U.S. Air Force, and Michael A. Fox, who wrote treatises in favor of animal testing—have done 180-degree turnarounds and now fight for animal rights. We shouldn't give up on anyone. The human mind is a very slow work in progress.

Q8. HEROES

Sojourner Truth, who was black, a woman, and basically illiterate and yet believed enough in her cause (the abolition of human slavery) that she stood up to heckling and physical abuse to make her points.

Dr. Neal Barnard, president of the Physicians Committee for Responsible Medicine (PCRM), who has given doctors a forum through which they can address animal issues, and who seems tireless in his own efforts to reform animal research and promote a vegan diet.

Peter Singer, whose sensible, plain, and moving book, *Animal Liberation*, often called the "bible of the animal rights movement," helped to congeal my thoughts about animals. When I read his chapter on the principles that bind social movements (the common theme that runs through them all), and put that together with his challenge to show how the other animals differ from us in ways that make it acceptable for us to eat or wear them or abuse them in other miserable ways, I changed from a welfarist who could still use them, to a liberationist who can no longer bear their oppression. To me, animals are not only feeling individuals who experience love and pain and hunger and loneliness and who want to live and to be happy as much as any of us does, they are also other nations with their own cultures. We may not understand them all the time, we may find their ways as odd as a Manhattanite finds a native of Borneo's ways odd, but we must respect them and their homeland rights. We have decimated the Native Americans who happen not to have the same arrogance that caused white settlers to decree that "the only good Injun is a dead Injun" and plant smallpox virus in blankets to kill them off so they could steal their lands. Today's equivalent is to bulldoze a small copse of trees—in which hundreds of diverse animals, from raccoons to frogs to birds, make their homes, never trashing or littering or ruining the environment, but preserving it—to putting up yet another mall or parking lot.

Q9. RELIGION/SPIRITUALITY

I believe that religions and spirituality require kindness, or they are worthless and their values lacking. All great religions were founded on the principal of kindness, but most people are religious in name only. Of course, wars and slavery and other horrors, including Nazism, have been supported by religions, too. That shows me that religion is a matter of human foibles and opinion. And then, no one knows if there is a god, so there is only blind faith. If there is a god, then he or she or it is either powerless to stop pain or has that power yet fails to use it to stop pain. That doesn't give anyone a reason to worship him/her/it. The only reason people worship a god they can't prove exists is out of fear that if they guess wrong, they'll be punished. You might as well believe in fairies. Recently, we have asked that the DNA on the shroud of Turin be tested to see if Jesus, or whoever was inside it, was a vegetarian. That should be interesting. Lots of people think Jesus belonged to a Jewish vegetarian sect, as we point out on our Web site (*www.Jesusveg.com*).

Q10. FUNNY/STRANGE EXPERIENCES

When 101 of us were occupying the fourth floor of a building at the National Institutes of Health in Bethesda, Maryland, and had been in position for about seventy-two hours, the *Today Show* called and wanted to interview me. I was able to sneak down a stairwell and out of the building without being detected, and appeared on the show. When I returned to the lobby, intending to give on-site support to the others still left upstairs, an officious security guard spotted me and took me back to the fourth floor, saying "I recognize you. You aren't supposed to be down here, are you?" It was incredible, but thanks to him I was able to resume my sit-in for another few days!

Another anecdote that I enjoy is when a police officer came to the PETA building to say that he had found a chicken sitting at a bus stop outside our office, and he wondered if it was ours. We thanked him for bringing us the new chicken, and he said, "I guess I should have realized that it wasn't your chicken. I guess you guys would always drive your chickens wherever they wanted to go."

Q11. LESSONS/ADVICE

Appreciate every single thing that anyone does, no matter how small, for the animals.
Don't worry what others may think about your actions and beliefs.
Don't be afraid to speak out and do what is right.
Be bold.
Be like the Energizer bunny, keep going.
Don't be intimidated by anyone or anything.
Never be depressed by the enormity of the situation. Look back and see how much has changed over the years.

Q12. ADDITIONAL INFORMATION

Read my books! There's *You Can Save the Animals: 251 Simple Ways to Stop Thoughtless Cruelty*, which actually contains more than seven hundred ways to help animals, from cooking vegetarian meals to starting a neighborhood watch for animals. It also has a great resource section, which is good for beginners, and tells you where to get free stuff. Its message is that everything you do and say helps. All activism, in the cause of fighting any injustice, is desperately needed. Then there's what I call my "cheap airport novel," a book called *Free the Animals!* which tells the story of how the animal liberation underground started in the United States. I've changed the names and disguised the identity of the people who lived those first adventures, freeing the animals, risking jail, and starting an animal rights revolution. It's an easy read. Finally, there are the PETA guides to compassionate living, vegetarianism, alternatives to dissection, leather-free companies, and health charities that do and don't test on animals. They are all available through our Web site (*www.peta-online.org*) or by writing or calling PETA in Norfolk, Virginia. We'd love to hear from any reader.

Ava Park

Q1. BIOGRAPHICAL PROFILE

In defining ourselves, we can use either "object referral" or "self-referral" methods. Most people, when asked "who they are," respond by way of object referral—external labels that identify themselves in relation to external things such as the name given to them by parents, what they are currently doing for a living, that sort of thing. Those things are all object referral, temporary and changeable, and not ultimately very important. Self-referral is how you know who you really

Ava Park. Photo credit: B. Dade

are down underneath, past all the external labels. The real answer to your question is "I am a developing spiritual entity currently inhabiting the plane of existence generally called Earth for the purpose of learning and growing."

If that's just too weird for everyone, then here is the more common version: I'm a forty-five-year-old single woman named Ava Park. In 1987, I founded a large nonprofit animal rights group, Orange County People for Animals (OCPA), and many people now help me run it. That is my volunteer work; I also run two businesses that serve to put groceries on the table.

I was born on February 22, 1955, in San Francisco. I enjoy studying metaphysics, philosophy, and spirituality. I love to travel, and particularly enjoy staying home alone from time to time, ignoring all my self-created obligations with the phone turned off, the front door locked, playing and talking with my four cats and one dog.

Q2. BECOMING INVOLVED

At age thirty, after a divorce, I went on a blind date with a doctor/researcher. We were headed out to dinner and a movie when he asked if we could stop by his experiment in progress at his lab at the University. He invited me into his lab, and the unspeakable horror I saw there changed my life and caused me to begin investigating what is done to helpless animals behind closed doors in laboratories for the privilege of publishing and profit.

Q3. IMPORTANT ISSUES

I spend much of my time making sure that the Orange County People for Animals continues to represent the animal rights point of view in our community. Because we are an animal rights group that addresses all issues, I don't spend too much time on any single given subject, but a little bit of time on all of them. Usually, when I speak in public, I discuss the topic of the "interconnectedness" among all beings. I have noticed that people have the ability to compartmentalize their minds and "put aside" issues that are uncomfortable for them to consider such as how animals actually got from the farm to their dinner plates. I continue to point out that this form of compartmentalization is unhealthy. We would all feel much better if we would integrate our behavior (what we eat, what we wear, how much we damage the planet ecologically) toward more ethical and honest activities.

The motto of OCPA, and my personal motto, is "creating a more compassionate, healthy and peaceful planet." I also feel that it is extremely important to bring to public attention the connection between human violence and violence to animals. This is another part of the "interconnectedness" theme that underlies most of what OCPA does.

I care so much about these issues because I believe it is critically important to the evolution of all life on the planet that human beings see and acknowledge their place in the circle of life. We are not above other forms of life. We are not above dogs, birds, snakes, cows, or fish. We are not above trees, water, rocks, or air. They each have a life force of their own that has intrinsic value, no less than our own. We are part of everything and must treat everything as precious. If we "use" other life forms (trees, water), which we are entitled to do to survive ourselves, we must do so with a sense of thoughtfulness, honor, respect, appreciation, and concern for future generations, not wanton short-term selfishness. Without this understanding and agreement among people all over the globe, we may ultimately destroy ourselves, our home called Earth, and all other life.

Q4. GOALS & STRATEGIES

My personal strategy is simply to be the best possible example of what I believe in. I also wish to continue enlisting the talents and abilities of our members to furthering our goals.

I write, research, educate, and speak publicly. Until recently I did a weekly live radio show called *Visionaries*, which interviewed leaders and authors in animal rights, human rights, environmental, and metaphysical and progressive human health issues.

I no longer organize protests because I believe in general that the day of the protest is coming to a close. There are better ways, I believe, in most (but not all!) instances, of achieving goals of change in society. Too many demonstrations are too confrontational, just providing an outlet for angry individuals to vent their rage, simply polarizing people, and not really creating any real permanent healthy change. There are better ways of persuading people to one's point of view than standing on a street corner, pissed off, holding a sign.

Q5. DEFINING SUCCESS

My personal ultimate goal is the same as the ultimate goal of OCPA . . . to have our Guiding Principles present in the fabric of everyday society. These principles are:

Ava Park

Every being has the right to live free from exploitation.
By educating people, we will create a compassionate, healthy and peaceful planet.
There are consequences to every individual's actions.
All life is interconnected.

(I would like at this point to give credit to OCPA's executive administrator, Veda Stram, for coming up with these principles and for being the driving force for making them identified with OCPA. All our members owe her a tremendous debt of gratitude.)

We will not "succeed" completely in my lifetime, or probably for many hundreds of years to come. But I can say that we have come a long, long way toward having these conversations take place within society, particularly here in Southern California. Laws being enacted currently and articles written in the press are just two pieces of evidence that show that the goals of OCPA and other animal rights activists are having a tremendous effect in changing our world. Also, our "Victories" list, which was paltry in its content just a few years ago, now must be updated often, reflecting frequent wonderful victories for us, for animals, and for all beings on the planet. All animal advocates should keep and look at such a Victory List regularly to keep their spirits up and encourage them to continue doing this lonely, difficult, and often very painfully sad work.

Q6. GROUPS & ORGANIZATIONS

Other than my own group, of course, Orange County People for Animals, I work with other wonderful groups in our area such as EarthSave and the Animal Rights Legislative Action Network. The OCPA supports their work by including them in our "Southern California Animal Rights Events" Monthly Calendar that is sent to our membership list. This calendar lists events for all groups that are taking place in our area. We also consider ourselves to be close to the Ark Trust (of which I am a Genesis Awards Committee Member), and People for the Ethical Treatment of Animals.

Q7. "THE OPPOSITION"

I prefer not to think in terms of black and white, of dualistic concepts. Humans are all on the same path, just at different places on the path. I believe that "that which you focus on expands," so thinking of anyone as "the opposition" simply puts more energy into them *being* the opposition and creating more "opposition" in one's life. I like to think of those with differing points of view—specifically, we are talking here about points of view that involve harming living beings—simply as people who have not yet awakened to the truth of their connection to all life. It is our job, the job of the members of OCPA, and the job I have chosen for myself in this lifetime, to help be a gentle wake-up call for them.

Q8. HEROES

I tremendously admire Ingrid Newkirk for creating the most powerful animal rights presence in the world. Having myself also tried to create an animal rights presence, I am very aware of how talented and brilliant she must have been in the early days, and must be now, to have accomplished this most challenging of tasks. I was privileged to be arrested with her at a demonstration against General Motors' animal testing in 1993. The time I spent talking quietly with her in the pokey of the Pasadena City Jail gave me even more reason to admire her and her personal power and determination.

I also admire, and am deeply grateful to the many animal advocates throughout history who were able to step out of the norms of their individual cultures (which often involved slavery and other abuses) and envision a better, completely different world . . . with perhaps no help or encouragement from those around them. These people include Thoreau, Voltaire, Gandhi, Lincoln, Schweitzer, Hippocrates, and Plutarch to name just a few.

Q9. RELIGION/SPIRITUALITY

Spirituality is not just part of my life, it is the underpinning upon which my life operates. I believe I could not do my work without it. It causes me considerable pain, however, because I am continually falling short of what I believe to be my best possible effort.

Q10. FUNNY/STRANGE EXPERIENCES

Well, animal rights doesn't lend itself to too many funny experiences, but there are plenty of strange ones!

We were the first group to take the carcass of a dead cow (this was several years ago, in our more confrontational stage!) and place it in the take-out line of a McDonald's fast food outlet so customers could "see" what they were ordering for lunch. Standing next to that poor dead cow and watching the many varied expressions on the faces of the McDonald's customers as they had to drive by him was certainly very strange for me, and I'm sure for all of them.

Q11. LESSONS/ADVICE

I would advise people new to the animal rights movement to lighten up already! So many new "converts" are so passionately indignant and angry, they end up alienating the very people they would persuade. (This was me during the first few years!) Be gentle with people who are just like you were before you found out all about it and became superior!

Be in it for the long haul. Too many people "burn their candle out" with intense work and energy in the first couple of years after they have learned about these outrages. But then they have nothing left for the long haul. That leaves the rest of us saying, "Hey, whatever happened to So-and-So?" Do what you have to in order to take care of yourself both emotionally and physically. The animal rights movement will take an unending, unlimited amount of your energy if you let it. Don't let it. Save something for yourself, or you eventually will have nothing left to give.

Let go of whatever guilt you may have for enjoying yourself while animals are suffering. Your being miserable won't alleviate their pain one iota. Do the best you can to reduce suffering by your own choices and by your own work to educate others. Then go have some fun, for heaven's sake. Frequently! Take the issue seriously, but don't take yourself seriously.

Don't get sucked into trick questions like "if it was just you starving and a deer on a desert island, would you kill the deer to eat?" or "if your child and a dog was drowning at the same time, and you could only save one, who would you choose?" These unanswerable hypotheticals are designed to do one thing and one thing only . . . to take attention and time away from the *real* topics that could be discussed such as "what are you going to have for lunch today?" and "are you going to buy a cloth coat or a fur coat?" These are real questions that affect real people's lives every day. Wasting time discussing silly hypotheticals is just a waste of time. Respond with the statement, "I'd love to discuss and debate with you *real* questions any time you're ready."

Be *for* something, not against something. That which you focus on always expands. This is a basic law of the universe. Let's be *for* compassion and gentleness, not *against* abuse and suffering. Stop giving energy to what you don't want. Give energy to that which you want more of, and you will attract more of that into your personal life and help build more of that on the planet in general.

Q12. ADDITIONAL INFORMATION

Believe it or not, no matter how it may appear, the universe is unfolding exactly as it should. Be at peace, and keep working for the evolution and happiness of all life, including yours.

Teresa Platt

Q1. BIOGRAPHICAL PROFILE

I was born in Canada the daughter of an agricultural banker, and raised in the farming and ranching communities of the plains of Canada and California's San Joaquin Valley. When my father transferred to Los Angeles to specialize in international finance for farming and fishing operations, I experienced the culture shock of moving from a rural area to one of the world's largest cities. My father switched to managing tuna fishing boats out of San Diego, and I worked for him part time. In 1986, I went to work for him full time when he purchased a fishing boat. I spent the next seven years managing shoreside operations for a two hundred-foot fishing boat and its eighteen-man crew.

I enjoy my family and friends, work, gardening, swimming, reading, and being around animals, the ocean, the desert, and the mountains, all of which are quickly accessible from my home in Southern California.

Q2. BECOMING INVOLVED

Animal rights/preservationist campaigns had a negative impact on the fishery I was involved in. The "dolphin safe" definition was shark- and tuna-deadly, negatively impacting a host of marine animals. It ignored the ecosystem approach to fisheries management. To counter this, I worked with the fishermen to start the Fishermen's Coalition, a nonprofit organization dedicated to educating the public about fisheries issues and holistic fisheries management. In the process, I learned that animal rightists were promoting diets and clothing from plants and

Teresa Platt. Photo credit: Simon Ward

synthetics, no matter what the environmental impact or the feasibility of the plan. Because less than 3 percent of the Earth can support crops to feed and clothe us, and because synthetics have major pollution problems, it is obvious to this country gal that although the animal rights philosophy may be a personal choice, it would make a poor public policy for 6 billion people sharing a finite planet.

Q3. IMPORTANT ISSUES

Everything I work on is related to the sustainable use of natural resources, and the humane use of animals. I now work for Fur Commission USA, which represents over six hundred mink- and fox-farming families on over four hundred farms in thirty-one states. Our volunteer board and committees work to ensure superior standards of animal husbandry through our own certification program, and to educate the public about responsible fur farming and the merits of fur. Farmed fur is a natural, environment-friendly resource. By feeding their livestock the waste from human food production, fur farmers reduce the environmental impact of the agricultural sector as a whole. And when a fur garment comes to the end of its long life, it's biodegradable too.

Fur farmers also make an important contribution to wildlife conservation. Farmed fur complements fur harvested as a part of wildlife management. By stabilizing prices in times of heavy demand, fur farmers help wildlife managers focus on ecological needs, not on market demands.

Fur is a superior insulator, a natural fiber that is durable, and exquisite to look at and to touch. Because more and more designers are incorporating real fur into their designs, educating designers, their clients, and the media on sustainable use helps us understand how humans are locked into an eternal symbiosis with all the other creatures on the planet.

I find the fur trade to be fascinating and extremely important. Because the only competition for fur for very cold weather is synthetics (petrochemical-based clothing), I see the choices for environmentally concerned consumers as obvious. For gentler weather, the choices include wool and leather, both animal products. Production of wool and leather results in "leftovers" not suitable for human consumption. These leftovers are fed, in turn, to other animals, carnivores, such as mink and fox. Fur production is part of the agricultural food and clothing chain with the farm-raised carnivores occupying a niche similar to the one they occupy in a wild setting.

Q4. GOALS & STRATEGIES

In a world where people are often forced to live disconnected from nature, I see education and a connection to the resource providers as key. I feel that it is important to provide people, particularly in urban areas, with the chance to talk with rural resource providers. I work with many other resource providers on bridging this gap using the new technology of the Internet.

I write regularly, which always requires much research. I think we resource providers not only have truth on our side, but we also have reality. Explaining reality is a challenge. I do work to educate people, and I have been involved in a number of protests. One of my favorites was organizing the San Diego–based tuna fishing families in the largest protest ever against Greenpeace. Eventually Greenpeace, along with four other major environmental groups, changed their policy, following the lead of the fishermen and the fishery's managers. That protest was peaceful and important in the process. I am proud to say I have never broken the law while protesting, and that

I have always focused my energies on moving issues forward through dialogue.

Q5. DEFINING SUCCESS

With tens of thousands of children dying of hunger daily, with inappropriate land use occurring globally, with too many chemicals and nonrenewable products used daily and billions more people expected to join us on planet Earth in my lifetime, we have some real challenges ahead. I believe resource providers have much knowledge to contribute to this debate, and ensuring that these vital people have a voice is very important to me. I believe we will have succeeded when human communities are healthy and functioning organically within healthy habitats.

Q6. GROUPS & ORGANIZATIONS

I work closely with many groups involved with the use of natural resources, ranging from local Farm Bureaus right up to national organizations such as American Agri-Women, the National Animal Interest Alliance, and the Alliance for America. I also work with many individuals involved in resource management, from locally elected representatives to state and national governments.

Q7. "THE OPPOSITION"

There is no opposition—just people who have not yet seen the light. Everyone is dependent on the Earth and its plants and animals. Some people like to pretend they are not, but with a little work, they can be made to understand the symbiosis between them and the Earth. Once they understand that they are dependent on the Earth and its plants and animals, then we can have a meaningful discussion on what the best methods are for feeding, clothing, and sheltering people.

I see the infusion of crime in the debate, intimidation tactics and even eco-terrorism, as a threat to all of us and to free speech. Arguments should never be won by force, they should be won by reason and I feel it is important that the outnumbered resource providers (who represent less than 2 percent of the U.S. population) not feel threatened when they express themselves. Resource providers have vast stores of knowledge to add to the debate and should be encouraged to participate.

Q8. HEROES

My world is full of heroes. They are farmers and fishermen and loggers and miners and medical researchers and designers and ranchers and sealers and whalers, hardworking good people.

One who really inspired me is Harold Medina who invented several designs for fishing nets, designs that helped release dolphins unharmed from the nets. Harold never patented the designs, he just gave them away for free. That is a true hero.

Bruce Vincent, a logger from Libby, Montana, is a hero to me. Bruce has a wealth of knowledge on forests and the animals that inhabit them as well as how to provide wood products sustainably.

I have always felt inspired by the steadfastness of Rose Comstock who has worked on the Quincy Library Group local timber management plan for many, many years against enormous odds from groups opposing local involvement.

I was inspired by the simple courage of farmers such as Tom and Carol Pipkorn of Escanaba, Michigan. After criminals invaded their fur farm in the middle of the night and stole and abandoned over five thousand of the farm animals, the Pipkorns worked tirelessly to recover the confused and stressed mink. They were helped by good neighbors (from far and wide), and thanks to those neighbors, the Pipkorns continue in animal husbandry today, providing mink pelts for clothing.

There are so many ordinary working people, resource providers, who I am proud to call my friends and heroes, I can't possibly name them all.

Q9. RELIGION/SPIRITUALITY

As one uncovers the Earth and all its secrets, one finds a revelation of the truth, which is God. It is humbling. Yes, spirituality is part of my life and helps me daily.

Q10. FUNNY/STRANGE EXPERIENCES

As a child, I was always bringing injured animals home and nursing them back to health. On a trip to a rocky beach in Canada, I found a slightly injured gull and carried it home back to a group of adults. One of the women couldn't look at the bird and turned away. She insisted we abandon the bird on the beach because she was too upset seeing it injured. She was too "sensitive," she said, to look at it. With a little care, this bird could have been saved so, as I left the bird on the beach knowing it would die slowly of starvation, I was confused by this woman's viewpoint. Just because one doesn't look at suffering, doesn't mean it doesn't exist. I believe people should accept that they are animals and dependent on the other animals.

As humans, we are the only animals to impose a code on our behavior toward other animals. To turn away is not "sensitive." It's an indication of the coldest of hearts. Since then I search for "hands on" people, not those promoting a "hands off" policy.

I am highly supportive of the conservation movement but run into many situations where this movement is distorted into an unregulated "conflict industry," continuously producing conflict for a buck, not promoting solutions. While I was involved in the eastern tropical Pacific Yellowfin tuna fishery, this conflict industry ignored the facts and provided highly inflammatory misinformation to the public in order to raise money on a charismatic mega-fauna, dolphin. They called this process "public information" but none of the information was put in perspective, resulting in a public that was terribly misled. Their own representatives told me that the public couldn't understand an ecosystem approach to managing fisheries and so they make the issue simple, focusing on one creature and promoting it over all others. Sharks, billfish, tuna, dozens of other creatures were suffering under the programs these groups were promoting, but they couldn't "sell" those creatures to the public and so these negatives were ignored. This single-species approach was deadly to holistic fisheries management, and the conflict groups knew this but simply continued the approach for financial gain. After a lot of soul-searching, several major environmental groups changed their policy to one of supporting the ecosystem approach in the eastern tropical Pacific, but they noted that this change would negatively impact their bottom line. This tells me that the groups that are acting as the ecological consciences of industry are driven by profit and that is a problem. Selling conflict is not a solution to the Earth's woes, and the public needs to be more cautious about which groups it supports; beware simplistic zero-tolerance solutions, wrapped up in hate propaganda against small groups of people one has never met. This approach will open wallets and raise donations but not solve real problems on the ground.

Q11. LESSONS/ADVICE

Make no assumptions and don't believe everything you read. Research an issue yourself and find the truth. Focus on what can be improved while remembering, respecting and including the people involved in the environmental equation. After all, we all depend on sustainable use of resources and the humane and sustainable use of animals. "No use" is not an option; it is nihilistic and suicidal.

And please join the resource providers in condemning acts of violence in this debate. If you could visit a fur farm after a "liberation" by animal rights terrorists, you would see something that looks like a M*A*S*H* unit with so many animals returned to the farm dead or near death. Dogs get many of the animals, as does stress and hunger, and tiny carcasses appear all over the countryside, bloated and decaying. Farmed mink associate the sound of motors with the sound of a feed cart, and this draws many of them to the highways where they meet their death, hopefully quickly. Days after an attack on a mink farm, when the feed cart is running at feed time, the "liberated" and hungry farm-raised mink will be seen circling the farm's fence, frantically trying to get back in. The farmer leaves the gates open at feeding time, so the starving mink can find their own way back in. As the mink wander home to the farm, some of them are in extremely poor shape. A mink will chew off its own tail rather than starve after misguided zealots "liberate" them. Among those mink that are recovered or make their own way home, the casualty rate is high. The mink herd will be put on antibiotics as a booster, but the casualties will mount. Some succumb to fights with their new "littermates."

Mink are born in litters of three to thirteen, and as they mature, they are broken down into groups of two or three littermates with which they then share a pen. When the "liberators" hit a farm, these littermates are forcibly separated. The farmer will attempt to pair the animals with new littermates, but sometimes they just don't get on. A fight breaks out and one mink ends up dead.

The dead animals littering the countryside after a forced "liberation" lived their entire lives on a farm, and were completely dependent on humans for their food and health. The "liberators" sacrificed these animals for a flawed philosophy that attempts to separate man from the Earth, an impossible notion.

Please join us in saying that animal rights terrorism is animal cruelty and should be condemned by all parties involved in the debate over man and his relationship to other animals.

Q12. ADDITIONAL INFORMATION

Find an issue close to you and where you live and give it at least ten years of your life. You can make a positive difference.

Susan Roghair

Q1. BIOGRAPHICAL PROFILE

I was born in Yorkshire, England, where the famous veterinarian James Herriott worked and wrote *All Creatures Great and Small.* My mother was British and my father French. My two brothers are engineers, and they still live in England. I also have relatives in England, Australia, and New Zealand. I lived in various parts of England (London, Bournemouth, Poole, and Surrey) before moving to the United States in 1986.

My hobbies and interests are so varied that they must seem like an odd mix for an animal rights activist. I write poetry and enjoy drawing and painting. I love interior design, especially American Victorian furniture. Making ceramics and dried flower arrangements, gardening, reading, and reviewing books for authors are a few of the other ways I spend my free time.

But my first passion is my love of animals. I share my home with Samuel Wedgewood Bartholomew, who is a thirteen-year-old Westie, and Molly Girl, who is a terrier mix about eight years old (she came from the shelter so I don't know her birthday). My other companions are Juicy Lucy, a little white poodle who came from the streets almost four years ago; Frankie Panky, a shelter rescue about seven years old; and a Cairn Terrier (just like Toto from *The Wizard of Oz*).

Q2. BECOMING INVOLVED

My first recollection of being an activist was at the age of about eleven, when I would stand in the city square of my hometown asking passersby to sign a petition to save the seals. I later became involved with protests that focused on fur issues and animal experimentation issues. My involvement with the animal rights movement has been a long process of growth. After settling in Tampa, I became involved with the local group Florida Voice for Animals, and then founded the Vegetarian Society of Tampa Bay. Other opportunities came along, such as becoming a senior chat room host for America Online for a weekly chat named *Animals and Society*. This led to another volunteer job, coordinating the speakers for a weekly national radio program called *Animal Forum* based in California.

Because of these activities, I have met many important leaders in the movement: authors, activists, and speakers. It has been exciting to meet so many people whose books I have read and whom I admire for their commitment.

The many connections I have made keep leading to new opportunities for me to help the animals.

Q3. IMPORTANT ISSUES

My heart stays constantly with the factory farmed animals, where most of the animal suffering goes on today. Each year in the United States alone, 9 billion animals are raised in pain and confinement and then killed for food.

Most people think that they need meat, but the irony is that animal products are the prime causes of this country's two biggest killers—heart disease and cancer.

Most people believe that farmed animals are treated humanely, when the truth is that animal welfare is the smallest concern of the animal food producers who are driven by maximum profits. I feel sorry for all the humans who are being misled by the greed of the meat and dairy industry and whose lives are cut short by wrong food choices. The facts are out there for all to see, in newspapers, magazines, broadcast media, and across the Internet—meat kills humans as well as animals.

Besides the enormous suffering of factory farmed animals, the meat-based diet is incredibly destructive of the environment. Before the end of this century, most humans will have been driven to a vegetarian diet because our natural resources will have been depleted. Half of this country's fresh water supply goes to livestock production, from growing animal feed to washing down the slaughterhouse floors. The U.S. meat industry is also responsible for a third of the fossil fuel used. We support the fishing industry to the point where it costs taxpayers many times the value of the fish caught to bring them to shore. It seems obvious that the meat industry is not sustainable in the long run. So I believe that people have an obligation to care for an inseparable triangle of concerns: their own health, the needless suffering of the animals, and the quality of the environment they leave for their children

I believe that once vegetarianism has become the norm, there will be no leather, fur, or other slaughterhouse by-products. Animals will no longer be used in medical research to find cures for diseases caused by eating animals.

Vegetarianism leads to feelings of empathy and compassion, so the cruelties of animal abuse for entertainment will disappear. I will continue to work on many other animal causes, but I feel helping people go vegetarian is a top concern because everything else will follow.

Q4. GOALS & STRATEGIES

Education is the key to making a difference. I believe that most people want to be compassionate, and that if I can let them know what is going on, they will try to stop participating in animal suffering. That's why for most of the last decade I have devoted so much of my time to online activism. My organization, Animal Rights Online, has grown to be the largest such group on the Web with almost five thousand e-mail contacts, individual activists, and animal rights organizations. My volunteer staff and I send to our e-mail subscribers weekly action alerts, book reviews, and two newsletters per week covering the latest in animal news, in-depth stories, vegan recipes, poetry, and other features.

Producing all of this requires a lot of time, but I am fortunate to have a great group of volunteers who help write and assemble the mailings and answer the large amount of daily e-mail questions that come in.

In the long term, I will continue to seize the opportunities that allow me to make a larger difference.

Q5. DEFINING SUCCESS

We will have succeeded when the needless suffering of all creatures has stopped. I know this won't happen in my lifetime, but I also know that I have played a part in leading us to a more compassionate world. Perhaps an end to needless suffering will come a bit sooner because of my efforts.

Q6. GROUPS & ORGANIZATIONS

Animal Rights Online works with many national and grassroots groups, and we believe that we help to make a lot of connections between groups. I was recently honored by being selected to join the Advisory Board of the Animal Rights Network, producers of *The Animals' Agenda* magazine, one of the best publications of its kind. I hope that this association between ARO and the ARN will be of mutual benefit. The more connected all of us in the movement can become, the more effective we can be in making meaningful changes.

Q7. "THE OPPOSITION"

I often think of the opposition as the people who make money from the exploitation of animals, the meat producers, the animal experimenters, and those who find enjoyment watching or participating in animal abuse for entertainment. But people can change, so I tend to think of people as potential converts and that the real enemies are Greed and Ignorance.

Q8. HEROES

There are so many people who have devoted their lives and made real differences for the animals that I hate to narrow my list of heroes down to two. But I think that John Robbins, who gave up his inheritance from the Baskin-Robbins empire to write books advocating veganism, has had an enormous impact on our movement. Another is Peter Singer, author of *Animal Liberation* and other books, who has raised the consciousness of a whole generation of compassionate people.

Q9. RELIGION/SPIRITUALITY

At times in the past, the "opposition" have somewhat scornfully remarked to me that animal rights or vegetarianism has become my religion. That bothered me a bit until I realized that there's nothing wrong with that. In her book, *The Inner Art of Vegetarianism*, Carol Adams describes how vegetarianism is a spiritual practice, and she creates a form of yoga based on it. Rynn Berry, the historical advisor to the North American Vegetarian Society, points out in his book, *Food for the Gods*, that respect for all creatures, the concept of Ahimsa, is the first precept of most of the world's major religions, and that vegetarianism logically follows this idea. Even organized Christianity, which has a long history of neglecting the rights of animals, supposedly had a vegetarian and animal rights activist as its founder, but according to Berry the early church suppressed these teachings of Jesus.

I also admire the writings of the Reverend Andrew Linzey of the Faculty of Theology at Oxford, whose books such as *Christianity and the Rights of Animals* and whose regular columns in *The Animals' Agenda* magazine have been inspirational for so many people.

Q10. FUNNY/STRANGE EXPERIENCES

Recently I participated in a "cattle liberation" at a New Mexican ranch, but as soon as we got the cattle to a safe pasture, UFOs abducted them all. I think that UFOs use animal methane for fuel.

Just kidding!

Q11. LESSONS/ADVICE

It's so important to take time for personal growth and happiness because you can't help the animals very much if you are hurting yourself. It is easy to be overwhelmed by the tremendous suffering of the animals and to lose sight of goals that can be achieved. Keep a steady pace, don't try to go full steam ahead, take time for yourself as well. Remember that there is a ripple effect that magnifies all our efforts; each person with whom you share the truth will go on to reveal that truth to others. It often happens that people feel ostracized and alone, and it helps so much to meet others who think like you. So join your local animal rights group or vegetarian society, make new friends in on-line chat rooms, cultivate relationships that help you grow.

Never forget that our objectives are reachable. We have made much progress in the last couple of decades even though there is still a long, long way to go. Howard Lyman once gave me some advice I'd like to pass along. Howard is a fourth-generation cattle rancher who sold his farm and became a leading author and speaker on veganism and Mad Cow disease. He once told me that people have to hear the truth about seven times before it begins to sink in, but once it sinks in you can never be the same again. Howard also believed that there was a turning point that would soon be reached: once a certain number of people had been made aware of vegetarianism, veganism, and animal rights, these ideas would become mainstream and there would be a fast conversion of the rest of the country. Thoughts like these keep me moving ahead.

Q12. ADDITIONAL INFORMATION

[No response.]

Anthony L. Rose

Q1. BIOGRAPHICAL PROFILE

I am Anthony Rose, called Tony by friends. I was born in Los Angeles in 1939 when this was a small city with big ideas. I grew up on the west side, mostly in Beverly Hills, and have lived most of my life in California. I have also traveled a good deal into Africa, Southeast Asia, Mexico, and Central America, and most recently to Australia. All my life I have enjoyed inquiry into the state of life about me, from the intricacies of humanity to the mysteries of nature. I like to hike and search for signs of biosynergy, sense the warm equatorial forest calming my nerves, sit on a boulder and look down across our family wilderness retreat in the high desert. My passion has turned increasingly toward the great apes, and I dearly love to be in the presence of gorillas, chimpanzees, bonobos, orangutans, gibbons, siamung. In truth, any primate will do—including human primates. We are after all the most fantastic of apes.

For nearly two decades, my most constant involvement with any living being has been with the California desert tortoises that have been breeding and burrowing in my backyard. I thought for much of that time that all we had in common was our will to live. Now I know that our underground ancestors have a power far stronger than the simple survival motive. Now I know that these ancient creatures carry out their long and fruitful lives with an unshakable and abiding faith. My involvement with tortoises is on their terms. They are in their element, their earth, and I watch awestruck by their resilience. To be buried and resurrected every year. To lay eggs without attachment. To eat and fast at the whim of time, at the rise and fall of the sun and prosperity. To die without fear. These are realities that I longed for but could not find as scientist. It took a fundamental shock to my humanity—the overturned stone, entrails, a shell-backed mentor's nod of farewell—to open me to the brilliant Earth life that only the reptile voice can articulate.

Q2. BECOMING INVOLVED

My first profound interspecies event occurred in 1963 while working as a research fellow at the UCLA Brain Research Institute. We young scientists were encouraged to experiment with anything, with soft furred rodents, with a lost dog, a litter of alley cats. Most of all we wanted close kin—chimpanzees. But we'd settle for monkeys: rhesus and nemestrina, cynomolgus, and the red faced Japanese macaque. Settle for distant cousins in gunmetal cages, a meter square; tree swingers boxed with barely room to stand.

I remember tough muscled monkeys pacing in small circles, rocking on red rumps, sucking thumbs. Hear them chatter high pitch at me in the morning when I troop into the lab. I inhale the thick brown musty smell of flaked skin, matted hair, and sawdust soaked in urine, caked to lumps of feces, burnt-bronze, texture of wet Purina monkey chow. I snicker back, smile my big teeth, huff thanks to turned backs, their shows of submission. If time allows I accept their invitations to put a forearm against the links, let brown fingers scratch for invisible ticks, let pink tongue lick salt from my skin. Those monkeys taught me how to talk with animals. For years we groomed and gestured, cooed and smacked. And for years I also restrained, invaded, implanted, and shocked.

Then one day a frightened janitor tracked me down in a colleague's office to tell me that a pigtailed macaque had escaped from his cage and was ransacking my lab. This had never happened to me before. Handling a scared monkey in a cage or experimental chamber was one thing: catching an escapee was another. I became terribly nervous as I walked through the long dingy corridor. I entered the room and peered through the haze and clutter; a familiar smacking sound drew my eyes to the far wall. Snicky, a three-year-old male, stared down at me from atop a bookcase, hair on end, eyes wide, teeth bared. Half terrified and thinking him hostile, unsure what to do, I mechanically smacked my lips at him, our usual morning greeting. He shuddered through a kind of tension meltdown and at once jumped from the shelf, leapt into my arms, and held on.

In the distance he had seemed so huge, imposing, wild. Now in my arms he was small, vulnerable, dependent. I sat on the linoleum floor with this animal in my lap—cleaned the scab that edged his dental cement skullcap, checked his implanted electrodes to be sure they hadn't loosened, and examined his dilated eyes. I remember thinking "after all I've done to him, he wants my friendship more than his freedom." I cried. This profound experience turned me away from medical research. I had become too bonded to continue. How could one experiment on his friends? I managed to complete my research and moved into a field where I could work with people in creative innovative endeavors—as far as I could get from the laboratory lives of trial and tragedy. I moved into applied social psychology and focused on human social change after getting my Ph.D. But I never forgot the monkeys. Then twenty years ago, I went to Sumatra to search for orangutan, and a bit later tracked after mountain gorillas in Rwanda. These natural epiphanies, compounded by my memories of those dear macaques, changed my worldview for good. Within a few years, I had turned my attention to great ape and rain forest conservation, and that is where I put my energy today.

Q3. IMPORTANT ISSUES

My main concerns have to do with our faulty worldviews that keep us focused only on human needs and human perspectives. This human-centered vision has stimulated the destruction of other animals and natural environments. It is the foundation for our own malaise, leaving us without awareness or contact with our origins and our homeland. We are in truth nature starved. Ironically we satisfy this deprivation by consuming nature, rather than nurturing ourselves in its living synergy.

Why do I care about other animals and natural ecosystems? Because I have allowed myself to commune with them. To become friend to the gorilla. To find home-space the rainforest. To acknowledge my own hominoid nature and teach my students to find their links to our once and living ancestors.

Why do I care? Because I have children who will inherit a world of concrete and smog and psychological deprivation if I do not care. Why do I care? Because there are gorillas in sanctuaries and in the forests of Africa who are dear to me as my own kin, and there are so many fine people, African and ex-patriot, who are working tirelessly to save those gorillas from destruction. I have committed to help keep those gorillas alive and support the work of those people. They have become my extended family and wherever they are is my homeland.

Q4. GOALS & STRATEGIES

I write, plan, fund-raise, lecture, organize, travel far and wide, facilitate teamwork, and build corporate and personal strategy—all aimed at protecting and understanding apes and their habitat.

Q5. DEFINING SUCCESS

Ultimately my goal is to enlist the involvement of the most powerful, wise, rich, and morally motivated people in the world to protect and defend the flora and fauna that lives in the wilderness—and to restore what has been destroyed. It is such people, united in this great pursuit, who will succeed.

Q6. GROUPS & ORGANIZATIONS

I consult with scores of animal welfare, conservation, and environmental action groups.

I am executive director of the Wildlife Protectors Fund, The Gorilla Foundation, The Biosynergy Institute, and the Bushmeat Project. See the Web site of the Bushmeat Project at *http://bushmeat.net*. I also participate as a steering committee member of the Bushmeat Crisis Task Force (*www.bushmeat.org*). I organized the Southern California Primate Research Forum, and I am a member of the IUCN/SSC African Primate Specialists Group.

Q7. "THE OPPOSITION"

There is no opposition. Only those who have not yet seen the wonder of nature and become committed to safeguard it.

Q8. HEROES

[No response.]

Q9. RELIGION/SPIRITUALITY

This is a spiritual quest for a very real salvation. We are not here to destroy Creation, to undo evolution, to civilize the planet—we are here to celebrate our vast potential for synergy with all life.

Q10. FUNNY/STRANGE EXPERIENCES

[No response.]

Q11. LESSONS/ADVICE

Study the so-called opposition and find what you have in common. Then work to unify your pursuits so that all succeed in furthering biosynergy—the synergy of life.

Our schemes and strivings to subdue and conquer nature are driven by fear. Scientist, teacher, politician, parent—all are afraid of being consumed in the potent exuberance of this elaborate and unknowable universe. Wilderness scares us. We hide from its truth.

We also hide its truth. A few years ago I talked to ethologists and conservationists at an international conference. My main thrust was to get them to loosen up, to report the personal aspects of their involvement with wild animals. When we describe how we face our fears and talk openly about our personal encounters with wildlife, we help the lay person to experience the scientist as a fellow human, and establish the groundwork for an informed sense of interspecies kinship. One scientist declared, "if taking a politician to meet a gorilla gets a vote for conservation, it's worth the risk of a little anthropomorphic thinking." Many agreed.

Still, a few argued that "we must never think of animals in human terms, it's unscientific." Ironically, this narrow-mindedness has undermined the natural sciences. Like all forms of ritualized simplicity, the law of parsimony and its correlated demand for severe detachment from and objectification of animals has hamstrung research, dehumanized animal caretaking, and dampened the spirit of conservation. Fortunately, recognition of the value of personal experience in these domains is growing. Empathy and intuition are again becoming legitimate factors in the understanding and support of other species. This allows us to attest to the secret truth of the scientific community—most people who study wildlife have experienced profound connections with animals and are deeply aware of the mysterious hidden lives of the animals.

Q12. ADDITIONAL INFORMATION

[No response.]

Andrew Rowan

Q1. BIOGRAPHICAL PROFILE

I was born in Bulawayo, Zimbabwe, in 1946. After two years in West Nicholson, Zimbabwe, a tiny community in the bushveld (baboons used to visit our garden and inspect me in my pram), the family then moved to Tristan da Cunha for three years. We then settled in Cape Town, South Africa, where I attended the University of Cape Town (studying Chemistry and Cell Physiology). After obtaining a B.Sc., I went to Oxford University where I obtained first a B.A. (Oxon.) in Biochemistry and then did a D.Phil., also in Biochemistry.

In 1976, I finally had to go out into "the big bad world" and earn a living. I worked for six months for Pergamon Press; then went to FRAME (Fund for the Replacement of Animals in Medical Experiments) as their first scientific administrator. My job at FRAME was to sell the idea of alternatives to the scientific community. The founder, Mrs. Dorothy Hegarty, did not want FRAME to be known as an animal welfare organization and insisted that we deal mostly with scientists and politicians and not "waste our time" attending animal protection meetings.

I learned a tremendous amount during my two-and-one-half years at FRAME, and then took a job with the Humane Society of the United States (HSUS) as associate director of the Institute for the Study of Animal Problems. In 1983, I started an almost fifteen-year stint with Tufts University School of Veterinary Medicine, where my task was to develop an "animals and society" program. When I left Tufts to return to the HSUS as a senior vice president, we had established the Center for Animals and Public Policy and had graduated two classes of students in the new (and unique in the world) Master of Science (animals and public policy) degree.

I have been an active sportsman, an avid reader, and a competitive bridge player (during my more dissolute university years). I now enjoy (mostly) raising three children, gardening, and exploring new approaches to animal welfare issues and challenges.

Q2. BECOMING INVOLVED

I became involved with animal issues when I took the job with FRAME in 1976. I had been raised by two scientists (one of whom was an ornithologist and naturalist) and was surrounded by nature and animals at home and on our family holidays. However, I applied for the job at FRAME not because of my interest in animals, but because I was interested in science policy and the FRAME position appeared to offer an interesting approach to dealing with that topic.

I had also seen FRAME literature (promoting alternatives to vivisection) while I was doing my doctorate, and remember being intrigued by the ideas presented therein. As I went through the interview process, I became more interested and more convinced that there were sound arguments to support the promotion of alternatives. Nothing that I have come across since has convinced me otherwise. While at FRAME, I began to delve more deeply into the animal protection literature and started to educate myself on the philosophy (beginning with Brigid Brophy's articles and rapidly moving on to Singer's *Animal Liberation*). The intellectual challenges drew me into the movement. However, I have also tended to side with the underdog in arguments, and animals are the ultimate underdogs in our modern world.

Q3. IMPORTANT ISSUES

I have always spent a considerable amount of time on the animal research issue. At FRAME, it was my whole job, and at the HSUS, it was still a major part of my responsibilities. These days I oversee animal research issues and am heavily involved in our initiative to eliminate pain and distress in animal research. For more than ten years, I had thought about setting up such a campaign and, on my return to the HSUS in 1997, finally had the means to do it. To me, animal pain, distress, and suffering are core issues for animal protection; and yet we have devoted remarkably little time to defining these concepts. We should develop techniques to determine when animals experience distress, and how much distress they are experiencing at any given moment. Our current campaign is dedicated to spurring others in the scientific community to address these issues and to devise ways to eliminate all significant animal distress from the laboratory.

I currently oversee the HSUS international animal protection program. One of the major challenges in promoting animal welfare in the developing world is that the humans frequently live in such poor conditions: only marginally better conditions than animals, in many cases. Under such circumstances, it is difficult to explain why one is trying to protect wildlife from trapping and snaring, when people nearby may be starving. We need to develop better justifications for promoting animal protection in these countries where human life is cheap, where human suffering is widespread, and where most humans have no day-to-day security.

Finally, I am very committed to promoting better scholarship and better analysis to support animal welfare positions in the developed world. The animal protection community has a tremendous weapon—namely, that the public in the industrialized nations expresses great concern for animal suffering. However, such concern is relatively shallow, and needs to be supported by analyses that back up this empathic commitment to animal protection with hard-nosed analyses of facts and values. The importance of Peter Singer's book (*Animal Liberation*) is that it provides a cogent and rational argument that explains why we should "feel" such concern for animals. In order to take advantage of the natural tendency for peoples' hearts to want to protect animals, we have to give their heads appropriate and supported reasons to do so.

In this regard, we are developing a broader educational initiative under the rubric of what we have termed Humane Society University (HSU). HSU is still very new, but we hope it will grow and expand to take care of the nurturing of human reason to follow in the footsteps of human hearts.

Q4. GOALS & STRATEGIES

I believe that we must harness academics more effectively to provide support for animal protection initiatives. We must also start to think more in terms of changing human behavior than in changing human attitudes. If we want to promote more sterilization of companion animals, we need to know what the current rate of sterilization is, why people do or do not have their companion animals sterilized, and what motivations we can use to get those who do not sterilize their animals to do so. We need to move away from the notion that we, as animal protection professionals, know what people "ought" to do; then think more of how we can get people to desire to act in ways that will promote animal welfare. We need to think in terms of getting "customers" to purchase our solutions rather than simply appealing to their consciences from the animal protection pulpit.

A few nonprofit organizations have been remarkably successful by thinking in terms of what the customer wants and then packaging their message so that it meets a perceived need of the "customer." My job at the moment consists mainly of managing programs. However, I do still try to do some teaching and lecturing, and I always try to keep up with new studies in animal welfare and animal protection. When a new idea comes along that seems to be useful (like social marketing), I try to disseminate it as widely as possible through networks of colleagues and staff.

Q5. DEFINING SUCCESS

I do not expect to reach my ultimate goal of a world free of violence toward both animals and humans. However, there are short-term goals. On the pain and distress initiative, I would like the USDA to establish a definition of distress (not just pain—that is a very different concept). They are in the process of doing this. I would like animal care professionals to be much more aggressive in looking for and eliminating animal distress. This process has begun, but it has a long way to go.

On humane education for children, I would like to identify what is currently being done around the country, so that we can evaluate whether it has a positive or neutral effect. I would like to understand what motivates children to be kind to animals, what motivates them to abuse animals, and then devise programs that encourage the one and lessen the other. We have plenty of humane education modules but very few have ever been evaluated to see if they actually change children's behavior.

Q6. GROUPS & ORGANIZATIONS

I work for the Humane Society of the United States and interact with a wide array of animal protection groups in the United States and internationally in the course of my daily duties.

Q7. "THE OPPOSITION"

There are a number of groups who would typically fall into an "opposition" camp. However, I believe that we make a mistake if we start to characterize the opposition as somehow morally bankrupt or evil. I follow the precepts set out by one of the most effective animal activists of all time—Henry Spira. He argued that we should not regard the opposition as bad, but should see them simply as individuals who have a different job to do and who see the world through different lenses than we do. If one approaches the opposition in this way, then it is easier to develop win-win solutions and to build effective multisector coalitions to support some action that achieves a specific advance for animal welfare.

The opposition groups for the animal research issue are typically such lobbying organizations as the National Association for Biomedical Research and the Federation of American Societies for Experimental Biology. In addition, establishment bodies such as the National Institutes of Health, and universities, are often opposed to actions taken to promote laboratory animal welfare. Industry occupies an interesting middle ground. Most consumer product and drug companies used to be "opponents," but, in the past decade (in part because of the work of Henry Spira), there is a much closer liaison between animal protection organizations and industrial companies.

In general, our "opposition" groups are either those who are engaged in using animals to produce a certain product for the market or who use animals as a source of entertainment.

Q8. HEROES

Henry Spira (unfortunately, recently deceased) was and still is somebody who has served as a major inspiration for my animal protection work. Other animal protectionists whom I admire include Henry Salt (nineteenth-century British author), Frederick "Doc" Thompsen of the United States (died in the mid-1970s), and Dr. Albert Leffingwell (died in 1916).

Q9. RELIGION/SPIRITUALITY

I do attend church but do not regard myself as a strongly religious person. Nevertheless, I do seek out writings on the topic and keep questioning my own beliefs to examine whether I am living a good and useful life. My religion does not affect my work much one way or the other. I would not describe myself as a spiritual person.

Q10. FUNNY/STRANGE EXPERIENCES

On one occasion, I was sitting at my desk when my boss came in and announced that we had to write and protest to the U.K. Department of the Environment about their plans to pour chemicals over the White Cliffs of Dover. When I asked for more information, I was told that the radio that morning had announced that an algae was growing on the chalk cliffs and turning them green and that the Department of the Environment was planning to remove the algae by pouring chemicals down the cliffs. When I suggested that the story might have something to do with the day's date—April 1—I was met with a blank stare. After further explanation, my boss could not stop talking about how radio stations wasted people's time!

Q11. LESSONS/ADVICE

I would suggest that any activist take the time and effort to learn what the opposition is saying and what arguments they are using to advance their case. It is quite common to find that the opposition's case is not necessarily watertight and to be able to come up with adequate counterarguments. If you know the opposition's case better than they do, it brings you credibility in front of any audience. The corollary is that you better be careful about putting forward claims that you have not adequately researched.

For example, I never looked into the animal activist claim (aimed at undermining the utility of animal studies) that "penicillin would not have been marketed if it had been tested on guinea pigs." Fortunately, I did not make such a claim in my book (but I have made the claim in talks). A few years ago, a researcher at the Research Defense Society pointed out that penicillin had been tested on guinea pigs (and the results published in 1943) but that penicillin had nevertheless been marketed for human use despite the mortality in the guinea pigs. What is more, the guinea pigs died as a result of a syndrome that occurred in some humans, so the guinea pig studies did serve to predict a relatively rare but significant side effect of penicillin in humans.

I have also been endlessly frustrated by the continuing use of the thalidomide disaster by animal activists as a criticism of animal testing. Most such critiques claim that thalidomide was adequately tested in animals prior to marketing. This was not the case, yet animal activists continue to repeat the claim because it endorses their own beliefs. As a critic of an establishment practice, one must be extra careful that one's criticism is accurate and valid. Even minor mistakes will damage one's credibility.

Q12. ADDITIONAL INFORMATION

[No response.]

Jerry Schill

Q1. BIOGRAPHICAL INFORMATION

I was born in Oil City, Pennsylvania, on March 21, 1948, and reared in a small farming/lumber community named Marble, Pennsylvania. My dad was postmaster of Marble, and the post office was in the front part of our house. I remember each spring when the local farmers would have their baby chicks delivered by mail. The boxes were about four feet square and six inches high with small holes throughout, where my brother and I would stick our fingers in and the chicks would peck at them.

    Although it was a two-story house, we would be awakened each morning by the muted sounds of activity downstairs when the carrier arrived with the mail between 6:00 and 6:30 A.M. I would often "help" Dad with mail sorting and postmarking envelopes. We also owned an eighty-acre farm a mile from town, where we raised cows, pigs, and chickens. We milked cows every single morning. Although my brother and I got out of that chore on school mornings, there was never a full day off. Not for Christmas, or any other day. Cows have to be milked twice daily.

    I attended Catholic parochial school, grades one through eight at St. Michael's Catholic parish, just a short walk from home. At that time, Benedictine nuns ran the school, and we attended Mass every day. The church is very large for a rural area, and sits as a dividing line on top of a hill between the towns of Marble and Fryburg. German Catholic immigrants settled the community in the early 1800s from the Black

Forest area of Germany, hence the name Fryburg. The huge, native stone, Gothic church was constructed in the 1890s.

Other than individual farms (many of them dairy farms), the largest employer was the lumber mill called Niederriter Lumber Company, owned by the Niederriter family. The family lived in the town of Marble, and I went to school for twelve years with one of the owner's sons. I can remember when the logs were skidded with Belgian draft horses and when the men would return from work in pickups retrofitted with homemade looking benches in the back. At that time, the mill was operated by steam. Coal was also big in that area, when I was growing up, and other family friends were part of the A.P. Weaver & Sons Coal Company in Fryburg.

I have two hobbies: photography and tractor restoration. I first became interested in photography in the late 1960s when Pam and I lived in Anchorage, Alaska, where I was stationed at Elmendorf AFB. We were married in 1968 after I had a tour of duty in Peshawar, West Pakistan, and we lived in Anchorage for two years until I was discharged in 1970. As a result of that experience, I opened a photography studio when we returned to Pennsylvania in 1970, and took photographs professionally for ten years. I have thousands of 35mm slides taken over the years. Many of them are family, but I have a substantial collection of scenics and landscapes, mostly of the Clarion County area of Pennsylvania. Recently I began photographing the commercial fishing industry in an effort to "put faces to the names" of the contentious battles that we're fighting.

I have a 1954 Farmall 400 tractor that I'm restoring in my garage. The tractor that I grew up with on our farm was a 1941 Farmall H, but this one will have to do for now.

Q2. BECOMING INVOLVED

I became involved through my present position, but strengthened by my life's experiences. I've been representing commercial fishermen since August 1987. It's only because the so-called animal rights activists have become interested in "fish" that I've been involved in this.

Interestingly, this issue may help commercial fishermen to mend the rift between themselves and sport fishermen. Because animal rights activists tend to vent their energies both toward those that "play" with fish and those that use fish for food, it allows both groups to have a common adversary.

Q3. IMPORTANT ISSUES

Very little of my time is devoted specifically toward animal rights activities. We forward information about their crazy antics to our members as a source of information and humor, but don't spend a lot of time doing it. As noted previously, we use animal rights information to show recreational and sport fishermen that they should be more worried about those crazies than commercial fishermen.

Q4. GOALS & STRATEGIES

First of all, it's important to note what our goals are here at the North Carolina Fisheries Association (NCFA). In nearly fifty years of representing the interests of our state's commercial fishermen, the goal has always been to assure that this historical and culturally significant way of life can continue. To that end, our goals are simply to make sure that fishermen's views are considered when new laws and/or regulations are being proposed, debated, and acted upon by various political entities and bureaucracies. Until recently, the "threat" posed by animal rights groups has been minimal to commercial fishermen. However, the threats by these groups are much more real to our political allies, that is, others in the United States who are involved in agriculture. Therefore, our immediate goal is to do whatever we can to aid those allies who are threatened by the radicals.

We know that our time is coming.

Q5. DEFINING SUCCESS

Because NCFA is nearly fifty years old, and I've been with the group nearly fourteen of those years, I think it's safe to say that we will never be able to say "we've succeeded." As long as there is money to be made by the radicals (green groups, animal rights groups, etc.), the pressure will always be on us. I do believe, however, that the pressure will be reduced significantly if our country ever gets back to understanding the importance of food production. Presently it's assumed that no matter what happens to farmers, fishermen, loggers, and so on, the products produced by those folks will always be at the grocery store. That's due to the disconnect that exists between the producers and the consumers. (Milk comes from the store, not a cow.)

Q6. GROUPS & ORGANIZATIONS

As previously noted, we work closely with groups such as the North Carolina Agribusiness Council and the Alliance for America. The Agribusiness Council is made up of various commodity groups that make up the food and fiber producers in our state. The Alliance is an umbrella group that looks at a multitude of issues in the property rights and environmental law areas, and attempts to put a human face and common sense into the debates. I have served on the Executive Board of the Alliance for America (AFA) for several years. We have taken busloads of folks to AFA's "Fly-In for Freedom," where the group participates in the discussion and demonstrations and can also meet with members of our congressional delegation.

The primary benefit for such activity is that commercial fishing families learn that they are not alone when it comes to surviving under burdensome regulations and radicals who threaten their livelihoods. They learn about logging communities that have been virtually shut down due to regulations, or fur farmers that have been victimized by ecoterrorists.

NCFA is also a member of the National Fisheries Institute and the National Federation of Independent Business, neither of which focus on the animal rights movement.

Q7. "THE OPPOSITION"

Although there are many animal rights groups in the United States, the one that seems to be the most visible in the fishing arena is People for the Ethical Treatment of Animals. That's due in part to their location, which is now Norfolk, Virginia. I should point out that I do not look at them as *the* opposition, but the most visible opposition within the animal rights groups. Our most visible opposition is the radical environmentalists and the organizations that portend to represent well-heeled sport fishermen. Interestingly, those sport fishing groups also view PETA as an adversary.

Q8. HEROES

Because I'm to the right of Jesse Helms politically, it should come as no surprise that I've always been an admirer of our senior senator, as well as former President Ronald Reagan. But regardless of ideology, I greatly admire anyone that has the courage to run for political office, and if successful, stay true to his or her convictions. When becoming weary of the battles and wondering why I remain in the fight, I can quickly recharge my batteries by remembering our founding fathers who not only risked their lives and fortunes, but in many cases forfeited them in the cause of freedom.

Q9. RELIGION/SPIRITUALITY

Religion (Christianity in my case) plays a very significant role in my efforts to represent commercial fishermen. First, many of the Apostles were commercial fishermen before being called by Christ. In the past I have been in charge of the Prayer Breakfast of the annual Fly-In for Freedom, sponsored by the Alliance for America, and frequently use biblical passages to encourage our participants to remain focused. I have been the vice chairman of our local Christian Coalition chapter for several years and am a member of the St. Paul Catholic Church.

Q10. FUNNY/STRANGE EXPERIENCES

A few years ago, our state's Marine Fisheries Commission held several hearings to get comments on some contentious proposed regulations. One of the hearings was held about three hundred miles inland, to give recreational fishermen an easy way to comment. I attended the hearing, and although I'm fairly well known on the coast as a commercial fishing spokesman, many to the west do not know me. When I spoke, I quoted a press release from PETA that was not too kind to recreational fishing or hunting. Because I did not begin my comments giving attribution to PETA, the audience thought the views were mine. I then introduced myself to the group and gave the proper attribution to PETA, and suggested that the fishermen should be more concerned about this radical organization instead of constantly berating commercial fishermen. After the hearing, I had some very pleasant discussions with these fishermen who agreed that we have some common ground.

Q11. LESSONS/ADVICE

[No response.]

Q12. ADDITIONAL INFORMATION

Commercial fishing is one of the last bastions of true free enterprise in the United States. The law of supply and demand still works in this business. It's interesting because you'll hear fishermen and fish dealers praise it and cuss it in the same breath. When a certain species first hits the market at high prices, there's nothing better. But when prices plummet due to high catches domestically, or if the markets are being filled by imports, then it's doom and gloom. Unlike many other industries, higher costs simply must be absorbed. Diesel fuel costs have doubled? Well, just tighten your belts, because if you don't sell at a certain price your customer will simply buy from a country that's subsidizing its fishing industry.

Although fishermen in some areas of our country are helped via tax dollars for buyouts and/or disaster assistance, North Carolina fishermen have been reluctant to accept such measures. Even with our recent endorsement of hurricane assistance for fishermen, some of our members felt that it was sheer folly. There are some who are normally opposed to any kind of handout, bailout, or government program. There are some who feel that if Ted Kennedy can get it for his fishermen, then why not us? But many just don't like the idea of taking anything from the "guvment."

In today's politically correct language, those women who opt to commercially fish insist on being called Commercial Fisher-MEN, not fisher-women. And for gosh sake, don't use the bureaucratic term of FISHERS with these hardworking men and women!

An economist type asked me if it would help commercial fishing if fishermen could have some sort of guaranteed annual income. My answer was that it would devastate, indeed ultimately destroy, that way of life. First, I told him to keep in mind that I have never personally been a commercial fisherman. Second, many fishermen may disagree. Nonetheless, I remain convinced of this. There are many facets to the allure of the life of commercial fishing. A lot of it we've heard before, and it is similar to that of many who work in the outdoors, especially rural folks. The fresh air, the freedom to come and go as I please (or at least that perception), the good feeling one gets from hard work, and so on and so forth. But the one element that so frightens most of us is what excites the commercial fishermen: the unknown.

Fishing vessels. What happens if my car quits running when I'm en route to a fish meeting? Well, maybe I'm late. What happens when a truck driver gets a flat tire? He pulls off the side of the road and calls for assistance. Other than the aggravation and lost time (money), he's no worse for wear. But if a fisherman is offshore and has engine problems, or gets gear caught in his wheel causing him to be dead in the water, then what? Depending on the weather, his very life and the lives of his crew are in jeopardy. That risk is true even without equipment failures, as the weather is always an unknown offshore, especially when working with booms, winches, cables, hooks, nets, and trawl doors.

Fish. If we're red-blooded Americans we like competition. If we're politicians, the competition is at the ballot box or on the floor debating a bill. Salesmen compete against their competitors across the street, or across the showroom floor. Even kids compete against their siblings for the attention of their parents. But a fisherman beats all. Regardless of how his fishing trip is, he's always concerned about how other fishermen have fared. Even if he has the best fishing trip he's ever experienced, somehow he just isn't "complete" if somebody else at the dock did him one better. That's not greed. That's being American! We didn't get to be number one by taking second best or being good. We got here by striving for number one, and commercial fishermen are very much a part of what makes us great.

Money. A commercial fisherman makes nary a penny unless he catches fish. When he leaves the dock, he is already in the negative with certain fixed expenses such as fuel. Depending on the fishery, a number of other fixed costs can be added, such as ice, bait, and groceries. In most cases, the owner of the fishing boat is also the operator. However, sometimes the owner hires someone else to operate the boat as its captain, and it's up to the captain to hire the crew. When the trip is over and the fish are packed at the dock, expenses for fuel, ice, groceries, bait, and so on, are deducted, and the captain and crew share the balance. These folks don't get a W-2 tax form at the end of the year; they get a 1099 form as an independent contractor. Again, they have no idea what their paycheck is going to be until they get back to the dock.

Most folks in today's society have a real hard time understanding the theory of getting paid according to production. Hence, the argument by some in government and academia that fishermen would view a guaranteed annual wage as a utopia. Not so! Academics aren't the only ones that have a hard time accepting this concept of risk. Bureaucrats cannot understand it, and even try to change it! Because they personally cannot imagine living such a life, they assume that guarantees are best for everyone. Guaranteed income isn't the only area either. In one of our successful court cases against the National Marine Fisheries Service, there was a very interesting exchange between Judge Robert Doumar and the federal attorney. In trying to understand the government's position, Judge Doumar asked the attorney if it was the government's intent to get rid of the highs and lows of fish populations so that it could always stay the same. The response was an astounding YES! The government not only wishes to play with the tried and proven free enterprise system with guarantees regarding economics, but also wishes to play God with fisheries management.

A fisherman has a love/hate relationship with this way of life. The allure of the water is that of the unknown. It is a spirit that relishes the risks on one hand, but despises them on another. It's a good life when a trip is culminated with a smooth running engine, gear that doesn't break down, good catches, and good prices, with a minimum of expense. That simply means a profitable trip for the owner, the captain, and the crew. But if the "unknown" is taken away from the experience, the allure is gone. A trip void of risk would be a boring trip indeed. Most of us sit on land thankful for the nice office, rolltop desk, weekly salary and perks, and expense accounts. A commercial fisherman would absolutely be destroyed in such a setting. His mental being is filled with what made this country what it is. A day filled with the unknown, both with safety and with economics. A pioneer spirit that is proud to be paid according to production rather than any sort of guarantee.

God help us when this country loses this last semblance of the pioneer in all of us! The guarantee that the fisherman does rightfully expect, is simply the right to work. The right to provide a healthy source of protein to a hungry world. But that's not what he's getting with today's fisheries management. He is stifled constantly by a barrage of regulation after regulation, with a new restriction being proposed before the ink is dry on the last. An Endangered Species Act that puts more importance on the protection of an animal than that of a human baby. An army of bureaucrats that has never known what it's like to be part of a private sector that pays according to production, and cannot in it's wildest dreams understand the economic risks, much less the real risk to one's life in pursuing a livelihood.

This army of bureaucrats does not understand that if one is to be able to experience the opportunity to succeed, one must also have the *opportunity to fail*. Guaranteed success only guarantees mediocrity. And mediocrity is not what made America what she is today.

Most bureaucrats love regulations, and decry fishermen for hating them. It's absolute nonsense to pose the argument that no regulations are needed in the fishing industry. When managing public trust resources, regulations are vital in the protection of these resources in the interests of the public who owns them, including consumers! However, for any politician or bureaucrat to expect fishermen to blindly accept them is sheer idiocy. The disdain that fishermen show toward regulations simply shows that they are red-blooded Americans, not the rogues that they're often painted by the extremists in government and environmental groups. The only American businessman that readily accepts regulations is the one that is trying to impede his competition, and that smacks at the very heart of the free enterprise system.

Cindy Schonholtz

Q1. BIOGRAPHICAL PROFILE

I was born in Marietta, Georgia, in November 1963 and have lived in Georgia, Mississippi, Louisiana, Alabama, Texas, California, and Colorado. I have a business degree from the University of Louisiana, Lafayette. I enjoy traveling and riding horses.

Q2. BECOMING INVOLVED

I have been involved in rodeo since high school. As a contestant and organizer of rodeos in the late 1980s and early 1990s, I began to be aware of the misinformation perpetuated by the animal rights movement about the care and management of rodeo livestock. I began to help on a volunteer basis writing letters, testifying at city council hearings, and educating the public. In 1997 I applied and was chosen as the animal welfare coordinator for the Professional Rodeo Cowboys Association. In my position, I am given the opportunity to clarify misinformation perpetuated by animal rights groups that wish to end the use of animals in entertainment, sport, and industry. I enjoy educating people about the care and treatment of rodeo animals and am motivated by the fact the people are so pleased to see rodeo livestock up close and learn that the animal rights groups are dead wrong about the life the animals lead, as well as their daily treatment.

Cindy Schonholtz.

Q3. IMPORTANT ISSUES

Education and legislation. It is important to educate the public about the truth regarding the treatment of rodeo livestock. It is also of the utmost importance that the rodeo industry not allow the propaganda of others to assist in passing laws that negatively affect the sport of rodeo or the use of animals. It amazes rodeo fans, contestants, and other rodeo proponents that animal rights activists will continuously spread lies about the sport of rodeo. Many rodeo committees, fans, contestants, and others are willing to help bring the truth to their communities through school programs, presentations to community groups, and other outlets.

Anyone who owns animals should be concerned about the vast amount of legislation that the animal rights movement is introducing to reduce animal owners' rights. Through the Professional Rodeo Cowboys Association (PRCA), the National Animal Interest Alliance, and the Colorado and American Horse Council, I work to defeat legislation that would negatively impact those who use animals in industry, entertainment, and sport. The amount of money animal rights groups will spend to pass legislation or get issues on the ballot is staggering, but they can and have been defeated in many of these issues. I see animal users growing more and more aware of the issues and endeavoring to educate themselves and others to assist in the fighting of negative legislation. I am able to reach many in the horse industry through a monthly column I author for *Western Horseman Magazine*, one of the oldest and most respected horse publications in the United States. The column focuses on legislative issues that the horse industry should address and reaches about five hundred thousand people.

Q4. GOALS & STRATEGIES

Short term—address current legislative issues; long term—proactively educate legislators. Invite them to rodeos and rodeo ranches to see the truth about rodeo livestock. In the education aspect, short term would be to distribute education information to grade school children about rodeo through local PRCA rodeo committees, as well as work on positive media about the sport of rodeo and the animals involved. Long-term methods would be to introduce legislation that hampers the terrorism and crimes of the animal rights movement and in the education area—and to have correct rodeo information available to all students.

I write for many publications including the *Pro Rodeo Sports News*, *Western Horseman Magazine, Friends of Rodeo Newsletter*, and others. I also am constantly researching through surveys conducted by on-site independent veterinarians at PRCA-sanctioned rodeos. These surveys allow me to compile figures of the rate of injury to rodeo livestock.

The latest study shows a rate of injury to rodeo livestock to be .0004. In addition, the PRCA is always looking at ways to improve upon our animal welfare programs and assist member rodeo committees in dealing with animal welfare and animal rights issues.

Q5. DEFINING SUCCESS

When the public and legislators understand the truth about the sport of rodeo and the treatment of the animals involved and when all rodeo associations go by the same standards for animal care.

Q6. GROUPS & ORGANIZATIONS

The PRCA works with almost all of the major rodeo associations and here are other organizations we work with:

American Association of Equine Practitioners—we network with equine veterinarians to learn the latest in equine health news and educate our members.

American Veterinary Medical Association—this organization assists the PRCA in educating veterinarians concerning the treatment of rodeo livestock.

Western Veterinary Conference—allows the PRCA to educate their member veterinarians about rodeo livestock welfare.

American Quarter Horse Association—the PRCA works closely with the AQHA on legislative issues as well as animal welfare guidelines for rodeo events.

National Animal Interest Alliance—this organization brings together those who utilize animals in entertainment, industry, and sport to address animal welfare and animal rights issues that affect all of them.

American Horse Council—the American Horse Council is the voice of the horse industry in Washington, D.C.; the PRCA networks with AHC to address any horse- or rodeo-related legislation or regulatory issues of concern.

Pikes Peak Humane Society—this is the local humane society in Colorado Springs. The PRCA works with them to educate humane officers concerning the handling of large animals.

National Conference of State Legislators—the PRCA attends this convention each year to educate legislators about the sport of rodeo.

American Humane Association (movie division)—although the American Humane Association has taken a position against rodeo events, following other animal rights groups, the movie division has worked with the PRCA. The PRCA assisted this division in creating standards for the filming of movies that include rodeo actions. Many PRCA rules were incorporated into these standards.

Q7. "THE OPPOSITION"

All animal rights organizations, the major ones being:

The Humane Society of the United States (HSUS)
People for the Ethical Treatment of Animals (PETA)
The American Society for the Prevention of Cruelty to Animals (ASPCA)
American Humane Association
In Defense of Animals

Q8. HEROES

I don't have an answer to this question.

Q9. RELIGION/SPIRITUALITY

Religion does not play much of a part in my work.

Q10. FUNNY/STRANGE EXPERIENCES

Every time I get involved in the political process, funny or strange situations arise. You never can second guess what is going to happen when dealing with politics!

Q11. LESSONS/ADVICE

For the person opposing animal rights activism, my advice would be to not get caught up in "fighting" the animal rights movement. Everything they do is to elicit emotions and try to get organizations and people to "fight." If you are fighting them, you do not have time to spend on proactive educational and legislative programs. Focus on your issues, not theirs. Work to improve the lives of animals, not fight those who wish to elevate their status and end their use.

Q12. ADDITIONAL INFORMATION

[No response.]

Mary Zeiss Stange

Q1. BIOGRAPHICAL PROFILE

I was born in Hackensack, New Jersey, in 1950—that makes me a baby boomer Jersey girl. I grew up in Rutherford, a middle-class town on the edge of the Meadowlands, when there were still some real meadows there, before they gave way to industrial parks and factory outlets. I believed everybody saw the Empire State Building when they looked to the east (even now some part of me regards New York City as the center of the known universe). And I believed a good day's hunting was best accomplished at the Garden State Plaza shopping mall in Paramus.

Still, I had girlhood fantasies of trick-riding with Annie Oakley, armed with my pearl-handled Dale Evans six-shooters, the ones that had come complete with a "silver"-tooled cartridge belt and a supply of red plastic bullets. Aside from those toy pistols, guns weren't a factor one way or the other, in my growing up. A member of the first TV generation, my action fantasies were shaped by Davy Crockett and Daniel Boone, *Wagon Train* and life on the Ponderosa. But if guns and frontier life were a natural part of my childhood fantasies, they were just as "naturally" foreign to the suburban New Jersey world in which I actually lived. I knew our next-door neighbors had guns; their two boys were scouts, and went camping and hunting. On my sixth birthday, the older boy had used the balloons strung up in our backyard for target practice with his BB gun. Twenty-five years passed before I came to understand how irresistible those targets must have looked to him!

Mary Zeiss Stange. Photo credit: Amy Berkley

When I was in my early thirties, I became intensely involved with both guns and hunting—I have written about this in my first book, *Woman the Hunter* (Boston: Beacon Press, 1998), at some length. Today, I would say that I am first and foremost a hunter—that is, hunting is not only the way I engage most intimately with the natural environment, but also the interpretive framework through which I tease out the meaning of that engagement. It is, in manifold ways, what keeps me alive, physically and spiritually.

I am also a committed feminist. Currently, I teach women's studies and religion courses at Skidmore College, where for eight years I directed the Women's Studies Program. And, with my husband Doug, I am a rancher; we operate the Crazy Woman Bison Ranch in southeastern Montana, where we focus on returning the buffalo to their historic range, as well as working with state and federal agencies to restore and conserve wildlife habitat. We are both quite committed to the model of the "hunter/environmentalist."

Q2. BECOMING INVOLVED

I suppose it was a gradual process—I had quit eating veal, and become concerned about animal welfare issues such as factory farming and cruel or redundant product testing, years before I began hunting. Indeed, from the start I saw hunting as a much more humane way of putting meat on the table than what goes on in feedlots or slaughterhouses. I also saw it as being on a direct continuum with my vegetable and herb gardening. It represented to me a more honest way of being personally involved with my food, with the sources of my sustenance, than anything I could buy shrink-wrapped in a supermarket. And I was, frankly, mystified by the vitriolic—and, I thought, quite simplistic—antihunting arguments touted by groups such as People for the Ethical Treatment of Animals and the Fund for Animals. Their arguments seemed as shrink-wrapped as the food I no longer found fit to eat. I was also insulted by their adoption of what purported to be some sort of moral high ground—as if their way of life didn't exact a toll on Mother Earth, in the same way that mine did.

And, as a feminist, I was particularly concerned about the way the animal rights movement, which on the grassroots level is about 80 percent female, identified women's and animals' suffering in such a way as to perpetuate the "victim feminist" idea that men are the bad guys and women, along with the nonhuman environment, are their inevitable victims and their moral superiors. I found—I still find—that idea logically inconsistent and ethically offensive.

Q3. IMPORTANT ISSUES

Well, hunting, obviously. And more broadly, the idea of ecofeminism. I do believe that there are some significant structural similarities between the subordination of women and human dominance over the nonhuman world. However, and again I have written about this at length in *Woman the Hunter*, I feel strongly that too much ecofeminist discourse, to date, misses the mark by oversimplifying the issues in such a way as to maintain a false polarity between bad male aggressors and good female victims. And this is especially true of ecofeminism grounded in animal rights activism. So my main focus has been on women hunters. It is a fact that our numbers are growing; there are arguably as many female hunters as there are female animal rights activists. What should we make of this fact? I think hunting, and environmentalism, and feminism, all look different—richer, fuller, if indeed more complicated—if we take the fact of women's hunting seriously into account. Given the depth and complexity of the social and environmental issues we currently face, I think we must get beyond the polarization that has characterized so much discourse about animal rights and nature rights. Focusing on women's reasons for hunting, and the ways they may go about it differently than many men do, is for me a good way to recontextualize the issues. This is what I'm up to, actually, with my coauthor Carol Oyster in our new book *Gun Women: Firearms and Feminism in Contemporary America* (New York: New York University Press, 2000).

Q4. GOALS & STRATEGIES

I am primarily a writer and an educator. For me the printed page and the classroom are the arenas for the free interplay of ideas. If we are going to build a better world, we need to start by expanding our base of knowledge on the one hand, while acknowledging the differing interests and perspectives that tend to make genuine conversation difficult on the other. Oftentimes, we evade those issues, which blocks real communication. Oftentimes, too, we write off people with whom we have ideological or cultural differences. In much of my writing—particularly in some editorial and journalistic work that I do for popular press venues like *USA Today*—I aim to create bridges for communication between disparate factions. I love it best when, for example, an ethical vegetarian and I walk away from a conversation feeling that we have actually reached some common ground. Or when a hard-core conservative "gun nut" reads a piece of mine on feminism and hunting and tells me it's the first time he's taken feminism seriously.

Q5. DEFINING SUCCESS

Frankly I wonder if "ultimate goals" aren't a part of the problem here. There are no comprehensive solutions for the social and environmental issues animal rights activists and their opponents generally address. The idea of ultimate goals has, for me, the nasty ring of "final solutions."

Feminist standpoint theory argues, persuasively, that there are no comprehensive solutions, in the sense that there are no solutions that serve all interests equally. So we all need, I think, to be prepared to brook a certain amount of compromise. Hunters, for example, can sacrifice certain practices that disturb animal rights activists—live pigeon shoots or long-range prairie dog shooting. And they can focus on "cleaning up their act" in terms of language and behavior, and on laying heavier emphasis (as female hunters, actually, seem to do) on outdoor ethics and sensitivity to environmental concerns. But for their part, animal rights activists need to stop demonizing hunters, and to confront the facts that most hunters do in fact care, many of them deeply, about animal welfare, and that hunting is a crucial part of wildlife management.

In any event, I would gauge "success" according to the degree to which the opposing sides can effect a meeting of minds that will really benefit the nonhuman environment; allowing us all to confront ourselves in the mirror in the morning and to live comfortably inside our own skins.

Q6. GROUPS & ORGANIZATIONS

I have been an instructor (of outdoor ethics, and large- and small-bore riflery) at a number of "Becoming an Outdoors-Woman" outdoor skills-training workshops for women. I belong to several women's and environmental organizations, but I am generally disinclined to speak or write on their behalf. Because of the controversial nature of much of my writing about women, guns, and hunting, I want people to accept my work as that of an independent thinker.

Q7. "THE OPPOSITION"

Those whose views are so intransigent that their minds are completely closed to arguments that challenge their own. I count a number of animal rights-oriented "vegetarian feminists" among them. But I also regard a number of died-in-the-wool good ol' boy hunters as "the opposition." Significantly, the vegetarian feminists and the ol' boys agree more with each other, about women in relation to men and to nonhuman nature, than they do with me—though neither side likes to be reminded of that!

In the spring of 1994, Friends of Animals (FoA) in conjunction with Feminists for Animal Rights (FAR) sponsored a conference in the Washington, D.C., area on the theme: "Eco-Visions: Women and Animals." Because the event was promoted as "a weekend of discussion and dialogue," and I had by then been thinking and writing about animal rights and hunting as women's issues for several years, I registered to participate. Two weeks before the conference, I received a phone call from an FoA official, informing me that my attendance would not be welcome. Subsequently, FoA's attorney, who characterized me as an intellectual fraud, contacted my college. The same attorney wrote me a letter warning that, should I attend the conference, I could face severe, albeit unspecified, consequences. All of this was, evidently, because I had dared to write articles critical of FAR and of the conference's keynote speaker. As it turns out, this keynote speaker later boasted that I had been "barred from" the conference. Ironically, until that point, I assumed that she had not been involved in the unfortunate incident; so much for sisterhood among scholars. My writings had simply pointed out a number of logical and conceptual inconsistencies in her thinking, none of which she deigned to address. As I wrote then, in response to her diatribe against me, "I would hardly use 'resentment' to characterize my feelings about that. 'Astonishment,' and a profound sense of embarrassment that this was done in the name of feminism, are more to the point."

As I said earlier, for me "the opposition" is those who refuse to accept the validity of any argument that challenges their own particular brand of orthodoxy. The "Eco-Visions" incident was a case in point. A former student of mine, who in 1994 was working for PETA, did attend the "Eco-Visions" conference, and said she was appalled that by the end of the weekend, "discussion and dialogue" had deteriorated into a shouting match. So perhaps I didn't miss much after all. But it deeply saddens me that FAR and other vegetarian feminist groups seem bent on perpetuating the divide-and-conquer approach to discourse that feminism, more generally, recognizes as part of the problem, not the solution.

Q8. HEROES

Annie Oakley, who knew that gender rules were made to be broken. Agnes Herbert, who hunted big game throughout the world in the late nineteenth century with her cousin Cecily, and wrote marvelously about it.

Q9. RELIGION/SPIRITUALITY

This is a rather more private question than I wish to address here.

Q10. FUNNY/STRANGE EXPERIENCES

This didn't happen in connection with activism, but it's a good illustration of the way assumptions about gender roles can affect perception: I was hunting pheasant in northeastern Montana, on the opening day of upland bird season several years ago. As is generally the case on opening weekend, game wardens were in the area checking licenses. I was stopped by a warden and gave him my Montana

Sportsman's license, which he perused for several minutes longer than I suppose either of us thought was necessary. Eventually he gave it back to me, and drove off. He clearly thought something was wrong, but he couldn't put his finger on it. My husband and I examined my license, and Doug saw the problem. Under "Sex," the clerk in the sporting goods store where I had bought my license had checked "Male." The clerk was so used to doing that, and the warden so accustomed to seeing the "M" checked off, that neither had consciously caught the error. My license, of course, was perfectly legal, but the warden knew something wasn't quite right about it. I'm not exactly ready for a fashion shoot after a day afield, but I am recognizably female. Whether the warden thought he was dealing with a diminutive man named Mary I'll never know. But in any event, I was statistically male for the 1986 Montana hunting season.

Q11. LESSONS/ADVICE

Suspect the validity of any argument that fits on a button or a bumper sticker. Recognize that there is a fundamental difference between sound bites and sound reasoning. Suspect any political or intellectual framework that casts individuals (e.g., males and females) or groups (e.g., liberals and conservatives) into relationships of polar opposition. Similarly, suspect any line of thinking that either erases distinctions between human and nonhuman on the one hand, or maintains rigid qualitative distinctions on the other. Be prepared to be disappointed in other people's capacity to let you down. But also be prepared for the occasional exhilarating shock of genuine connection that occurs when, sometimes through an arduous argumentative process but just as often as if by accident, you and an ideological opponent reach a position of mutual understanding and respect.

Because it is in those moments that real communication begins.

Q12. ADDITIONAL INFORMATION

[No response.]

Patti Strand

Q1. BIOGRAPHICAL PROFILE

I was born in 1946 in New York City. Shortly after Rod and I married in 1970, we brought two dalmations into our home. We raised them and became involved in the organized dog world, now for about thirty years.

Q2. BECOMING INVOLVED

It was a gradual process for me. It started when friends asked me to organize support for a piece of state legislation ostensibly drafted to fight "puppy mills," the substandard dog breeding kennels that all conscientious dog breeders abhor. I had no previous legislative experience. When I actually read the bill, I found that the language was so broadly drafted that its passage would have prevented even the most responsible people from being able to legally breed a litter. For instance, it required more kennel space for a dog than what was required to house children in day care centers. Over the course of the legislative session, I met several people associated with farming, hunting, and biomedical research, who were likewise engaged in opposing extreme proposals that would affect their avocations and businesses.

They proved to me that the people behind the so-called "puppy-mill bill" were zealots who espoused a philosophy opposed to all uses of animals, no matter how responsible, how humane, or how critically important. They also showed me that we were fighting exactly the same people.

By now I realized that I was dealing with unethical people, but I wasn't ready to buy into the whole theory. Over the course of the legislative session, however, it became clear that the bill's supporters were not interested in compromise; they were willing to do whatever it took in order to get their bill passed. The bill's nominal drafter was an out-of-state resident until just days before the legislative session began. Incredible misrepresentations were made to sway legislators to pass the bill. I was disturbed by their willingness to subvert the process. I was naive about both the animal rights movement and about modern legislative processes. Before the session ended, however, an ALF (Animal Liberation Front) bombing at Oregon State University opened my mind to the possibility that indeed we were encountering organized extremists.

Following the legislative session and the ultimate demise of the bill, I wrote a few articles for dog publications. Although I did not mention anyone or any group by name in my early articles, I was very critical of the zealotry of the animal rights activists whom I had encountered, and of the misinformation campaign they promoted. I became an activist when, immediately following the publication of my first article, I began receiving threatening hate mail (lots of it). It was apparent that the letter-writing campaign was orchestrated, if not by the leaders of the movement, at least with their approval. It was either directed at me specifically, or perhaps I was on a list of targeted people singled out for intimidation. The warnings were graphic, and directed at my family members, as well as myself.

Q3. IMPORTANT ISSUES

I work continuously to expose the misinformation campaigns of the animal rights movement. I believe that the style of operation used by the animal rightists undermines basic tenets of a free society. This style exploits, it does not serve, the welfare of the animals that they profess to care about. I believe that the leadership is corrupt and misuses the compassion of the followers, whom they attract through misrepresentations. I think that the movement is a cult that misuses its followers.

People should care because at its core the animal rights movement is a hate movement. It inflames people to hate certain people, to sabotage industries, and to vilify individuals through the use of distortions, half-truths, outright lies, and sensationalized film footage. It encourages violence done in the service of their avowed cause. By misusing our rights to assembly, free speech, and so on, they subvert the foundations of freedom. The movement is a fraud. They are true animal exploiters.

Q4. GOALS & STRATEGIES

Our goal is to inform the American public about the corruption in the animal rights movement. We distribute articles and press releases that expose the fraud, extortion, and criminal support found in the animal rights movement.

Our Web site has won a Reader's Digest Editors Choice Award for content (*www.naiaonline.org*). We publish a quarterly newspaper, and many of our members regularly publish information in trade and scientific journals about animal issues and the animal rights movement.

We maintain files and a database of factual information about animal issues, along with events and individuals in the animal rights movement. We seek to educate people by distributing our information, and we serve advisory boards that develop public policies on animal issues.

Q5. DEFINING SUCCESS

I will feel successful when the criminal element of the animal rights movement is behind bars, and when the movement is led by ethical people.

Q6. GROUPS & ORGANIZATIONS

The National Animal Interests Alliance works with any group that is working responsibly and within the law to improve the welfare of animals and eliminate extremism within the animal rights movement.

Q7. "THE OPPOSITION"

Among many others, I consider People for the Ethical Treatment of Animals (PETA), The Humane Society of the United States, Fund for Animals, The Sea Shepherd Society, and Coalition to abolish the Fur Trade to be groups that operate unethically (and in some cases, illegally).

Q8. HEROES

Martin Luther King, Abraham Lincoln, and Aristotle are some of my heroes.

Q9. RELIGION/SPIRITUALITY

I am a religious person in a nondenominational, non-churchgoing way. I am motivated more by compassion for my fellow man, and a commitment to promoting a civil and just society, than I am by religion.

Q10. FUNNY/STRANGE EXPERIENCES

A phone call from Ingrid Newkirk aimed at intimidating me, when I first started writing about the movement, was an extremely strange experience.

Q11. LESSONS/ADVICE

I would have to write a book to answer this question fully. The short answer is a paraphrase: "The only thing necessary for evil to prevail is for good people to do nothing."

Q12. ADDITIONAL INFORMATION

When demonization of others is required in order for any cause to succeed, the cause should be questioned. People should research information carefully and make sure that it is not based on destructive half-truths. Hate movements, whether doing business under the cloak of serving white supremacy, unborn babies, or animal rights, are in the end still hate movements. People who want to help animals can make better use of their time, energy, and money by investing in local shelters, 4-H groups, or by supporting rescue and education, rather than by attacking people and industries.

Michael Tobias

Q1. BIOGRAPHICAL PROFILE

Of Russian/Jewish descent, I first appeared in San Francisco, according to my parents. I only vaguely recall the day. I have, by turns, enjoyed and suffered ever since. This has molded my character and outlook. I must assume that human nature is given to these sharp twists of perception that make for great moments of precious abandon or oblivion, and other moments utterly shaped by the acute recognition of pain and suffering throughout the world. From my first stirrings I vividly recall asking the same question: Why must there be suffering in the world? Might it not be avoided?

Q2. BECOMING INVOLVED

As to the above referenced question, about suffering and how to remedy its impact . . . I recall my first encounter with a caged wolf at the San Francisco zoo. I must have been two or three years old. That enraged, frustrated, desperate animal gazed upon me through the bars condemning it to purgatory with the most memorably forlorn eyes; eyes, groping questions, incredulity that even today remain with me at the core of my lonely horror. Later on, as my personal indignation at the behavior of my fellow species grew, environmental activism was simply the only option to me as one endowed with a conscience. I grew up with companion animals (we called them "pets" then)—cats and dogs, turtles and fish. By my early teens, I was absorbed by philosophical readings, particularly Albert Schweitzer, his cousin Jean Paul Sartre, Albert Camus, Hegel, Marx, Schopenhauer, Samuel Beckett, Joyce,

Michael Tobias.

and so on. Painters and musicians were equally important to me: Monteverdi, Bach, Shostakovich, Kuo Hsi, Giorgione, George Inness, Vermeer, and Van Eyck. My mother, who was, among many other things, a great painter, also helped me to see nature, and to appreciate the interdependency of all things. My father, an engineer and lover of Shakespeare, impressed upon me an additional component to the philosophical traditions, namely, the human need for a livelihood. What took shape in my thinking was this tragedy of human enterprise and the tide of destruction resulting from our best pursuits and interests. As a species, we were all somehow connected to this vast web of impact and assault. Loving families, artists, thinkers—each of us additionally burdening the natural world with our own drives and ambitions. I could not help but recognize the end-losers in this tide of being, namely, other animals and plants and all of the ecosystems which, in the 1960s, were just coming into a kind of national "vogue" of interest. Rachel Carson and Peter Matthiessen had written their astounding books on pollution and wildlife decimation. I read these with profound dismay and horror. By then, my path was clear. I was also at that point familiarizing myself with Thomas Malthus on population, and Jeremy Bentham and Tibetan Buddhism on suffering. Ecosystem studies were becoming key to biology departments. Protests over vivisection became more and more common. The Vietnam War might have distracted many from the plight of ecology, but some of us were questioning the demise of charismatic megafauna in Vietnam, as well as people. We were in a minority, but the "movement" was growing. Of course, it had its ample precedents, down through the ages, and those who had read Clarence Glacken's monumental *Traces of the Rhodian Shore* were more than vindicated. There were Thales, Lucretius, Aristotle, and countless others throughout time who had studied and voiced the rights of nature, the beauty of nature, the crucial importance of wilderness. By this time I knew without hesitation that human survival depended upon our ability to love and cherish the natural world.

Q3. IMPORTANT ISSUES

I am deeply concerned about so-called "biodiversity" loss and the infliction of pain on other beings. My wife, Jane, and I are involved in the rescue and protection of domestic and wild species and habitat, but we never generalize habitat to mean anything other than the vast number of individuals who are vulnerable and suffer as a result of human greed, ignorance, and outright malevolence. We are especially concerned about so-called "farm animals" who suffer in the tens-of-billions every year (tortured, killed, and consumed by humans). What does torture really mean? The annals of human history have documented the verb/noun to accord with the common understanding of murder, sadistic infliction of pain, and genocide. These are the appropriate words to describe the condition to which most animals are subjected in all the world's farms. Farmers who may lavish, in their minds, great affection upon their animals, think nothing of slaughtering them when the day comes to earn some money. There is a large debate in England presently about the "plight" of small slaughterhouses that are being put out of business by the U.K. agricultural provisions that mandate the presence of a veterinarian at the killing of every animal. The farmers argue that they cannot afford such scrutiny. Moreover, Trust lands in the United Kingdom maintain many farms where slaughter is done according to so-called "organic" standards. All such standards will probably be "improved" as a result of the foot-and-mouth madness—the absolutely unnecessary slaughter of hundreds of thousands of animals in order to maintain the veneer of "100% clean meat." It is hard to envision cleanliness amid competing images in our minds—and the reality on the farm—of an animal holocaust that suddenly swept the countryside for no economic reason. And in truth, the holocaust has been happening all along, not just on the farms, of course, but in laboratories, on roadsides, in industry and government. Animals are killed for human use and little if any notion of their moral rights or rich cognitive and emotional lives ever surfaces or is granted legitimacy by the human species. It is now estimated, for example, that 100 million sharks per year are being slaughtered, mostly for their fins. Millions of animals—and more and more genetically modified ones—are tested in laboratories where it is increasingly difficult to separate medical altruism (i.e., empathy) from the profit motive. In the United States, slaughterers would like to use the term "organic" for much of their own killing. But the word organic is traditionally defined (in the Oxford English Dictionary) as connoting a musical air, something pastoral and fine, loving and good. The perversion of the word, that is, the overall subjection of animals to human consumption, all amounts to a vast trespass by one species over another, farm animals. This trespass then continues in less easily defined ways, across all the habitat of the accessible world. Even the inaccessible places are affected (drifting pollution, dead zones, radiation, and lead deposition as far away as Greenland and the Antarctic). Out of the last eight years, pack ice only formed twice, around the Southern Continent—a direct and unambiguous result of Global Warming, sounding a certain death knell for phytoplankton, krill, penguins, leopard seals and whales—an ecological crash that is likely to be sudden and unprecedented, notwithstanding a political administration (in the United States) which prefers to take its merry time studying the situation. We have usurped more than half of the terrestrial and marine worlds. Biologists speak in terms of NPP—the net primary production, or chlorophyll-producing areas, that have been overrun by humans. The resulting species loss is nearly incalculable, but most biologists who take time to examine these issues concur that at current rates we will lose nearly 70 percent of all life on earth in coming generations. That translates, crudely, into quadrillions of organisms that we will, in essence, kill by our indifference. It is likely, given the current demographic winter which is not expected to get much better, that this grim scenario will take place much like a colliding asteroid. We have unleashed the sixth extinction spasm in the history of life on earth and it is at least 20,000 times more deadly than the natural background rate of extinctions. Given that we number over six billion and are poised to double as a species, our collective impact is overwhelming. By direct and indirect assault we are killing everything.

Yet, I believe there is no limit to the human heart. Focusing its capacity on specific issues of habitat conservation and the amelioration of suffering among individuals (whether you call them "domestic farm animals" or wild individuals) is what my life—and that of my wife Jane—is devoted to. I do not believe that family planners will stop the population bomb. But I believe that 12 billion ungainly, bipedal largely carnivorous Homo sapiens can—if gently persuaded—choose to behave themselves, to act like mature, rational, empathetic beings, as opposed to monsters.

Q4. GOALS & STRATEGIES

Jane and I have been involved in environmental work over the years in dozens of countries. Our short-term goals involve saving as many individuals and as much precious habitat as possible. We both write, direct, and produce films and books and articles on this topic.

I also teach (in the past, at such universities as Dartmouth, California State University-Northridge, the University of New Mexico and the University of California-Santa Barbara), conduct multidisciplinary research, and continue to travel to the front lines in many locales, never ignoring my backyard. I try as best I can to observe, listen, interact, and insinuate my own beliefs where I feel they may have a chance at affecting those of other persuasions. The day only holds so many hours, but a striking image—a forcefully positioned point of view, articulated well, can transcend its physical barriers in time and space. My films are difficult to exhume. Once they air, they exist in network vaults. Some are available from The Video Project in Ben Lomand, California; and others from Turner Broadcasting, PBS, Discovery Channel, and Discovery's Animal Planet network, as well as PCI—Population Communication, Inc. A recent movie for ABC (July 15, 2000), *The Sky's On Fire*, which I co-executive produced and which was based upon my novel, *Fatal Exposure*, was about ozone depletion. My docudrama, *River of Love*, filmed in southern India, can be found on *www.mysticfire.com*. As for my books (there are twenty-five available), most can be found at *www.abe.com*; *www.bookfinder.com*; *www.amazon.com*; and *www.bn.com*. One Website lists many of my works: *www.rationalskies.com/youthtopias/tobias/tobias.htm*.

Q5. DEFINING SUCCESS

Every time we save an animal, or help an existing organization to effect concrete changes in the status of imperiled animals and plants, we have succeeded. My ultimate goal is to see a bill in Congress ending meat eating and the vast "incidental take" of species, endangered or otherwise; providing legal status to all species, in other words. With Mad-Cow Disease and others of its ilk, and all the ills associated with the killing of other animals to serve ourselves, I see such a bill looming large. There will be difficult debates as to how 12 billion humans can learn to walk more softly, without stepping on the ants, and murdering nearly every other species we can get our hands on, but those debates must ensue.

Q6. GROUPS & ORGANIZATIONS

Conservation International in Washington has directly or indirectly saved over 80 million acres worldwide—a phenomenal model of biological activism.

PETA, People for the Ethical Treatment of Animals has proved to be extraordinarily effective at raising debate to levels of legitimacy, and impacting positively on pernicious corporate and governmental forms of animal cruelty and indifference. PETA's president (Ingrid Newkirk) and I, among others, were in India some time ago to try to gain and reveal information that might help PETA curb the cruelty inflicted on cows in that country. India tortures its animals while doing a public relations job on the rest of the world that would have non-Indians believe that animals in that country are sacred; the cow revered. In truth, cows are tortured and slaughtered in India, just like everywhere else. There are approximately 100 million vegetarians in India (10% of the population) but even that large force of people have not aligned themselves—as they must—into a political rallying cry. They have neither stopped the abuse of animals, or of children, or of women. Nonetheless, what PETA has done in the aftermath of our visit has been to get the issue out into the world press. I know that such efforts have engendered some good. Indian leather manufacturers—and corporate importers in the West—are now concerned about their image abroad and this might dampen their zeal to slaughter animals for purses and shoes. Farm Sanctuary has enlightened countless people about the suffering that occurs in factory and other types of farming. There are, in addition, many other environmental and animal rights organizations, in many countries, that are doing this kind of work. All of them need public support. Desperately so.

Q7. "THE OPPOSITION"

I don't ascribe to the notion of "opposition." We are all born of the same blood and basic genetic disposition as a species. Seen from space, we are one massive organism. When you argue with a loved one or a friend, are they the opposition? No. We all wake up, at times, on the wrong side of the perch. There are patterns of consumption and callous indifference that have emerged over tens-of-thousands of years within the human species. Those patterns are inappropriate and must be halted. Every activist knows what I mean. Patterns of sorrowful, unnecessary infliction of pain and destruction. But I do not believe that direct conflict is the most effective means of halting such harm. Nor do I ascribe to passive meditation. Both "techniques" may be appropriate to certain individuals. My observations tell me that there is little time left to save this biological house of cards upon which all remaining life forms depend. We must act, unambiguously. But not by perpetuating conflict and disarray. Our words and actions must be measured, temperate, balanced. We had to bomb Hitler's Germany and we did not act swiftly enough to save most of Europe's Jews and other victims (though some were saved). But we cannot approach the animal holocaust with similar aerial sorties, or hand-to-hand combat, using weapons of destruction. We must engage those with differing viewpoints through nonviolence and informed example. We need to become beacons of an alternative approach: lighthouses for the new way. I refer to the engendering of a new human nature, and each of us who thinks about these things has the responsibility to devise ingenious methods for the cause of liberating all sentient beings. This is very much a part of the religious belief known as Jainism. But it is also very Jewish and Christian and Islamic and Buddhist and . . . such thinking comports with every spiritual or tribal ethic of which I am aware, everywhere in the world. The techniques for consensus building, regarding compassion and alternative economic solutions (for all social and economic strata), must emerge through a balance of social justice and ecological justice. Six billion consumers constitute an enormous challenge to policy and lawmakers, and to the public itself. But we absolutely must provide legal voice, dignity, and "personhood" (as Steven Wise uses the term) to each living being. I don't mean just primates. I also mean ants and wasps and bats and boll weevils. Pigs, turkeys, sheep, horses, tigers, and rhinos. There are several million species left on Earth, and hundreds of trillions of individuals. Should every microbe have the same rights as you or me? Kill my microbes (i.e., with an overdose of antibiotics) and you might kill me. You see? The ethics do not break apart across species boundaries. But the practical choices are real, infuriating, and heartbreaking. We have to engage in triage. It is inherent to nature and evolution. And this is where the debate gets terribly confusing for all of us. My model here comes from the Jain tradition, to which I am strongly affiliated. Each individual activist is challenged to devise techniques that will be effective, not just loud, or busy. Effective means compassion in practice. It ultimately translates into minimizing violence. You cannot stop it.

Q8. HEROES

I am deeply impressed by Jane Gray Morrison, a great animal rights activist and the most deeply empathetic human being I have ever encountered. I am fortunate indeed to be her husband.

There are others who have also inspired me greatly. Mahavira, the 24th Jain sage (a contemporary of Buddha). Lao Tzu, Han Shan, St. Francis, Leonardo da Vinci, Percy Shelley, Nikos Kazantzakis, Peter Matthiessen, Ingrid Newkirk, Gene and Lori Bauston, Marc Bekoff, such other painters as Paulus Potter, Frans van Mieris the Elder, Filippo Lippi and the French Intimists, Kimon Friar, Robert Radin, Russell Mittermeier (Conservation International), Norman Myers, Thomas Lovejoy, Henry Purcell, John Muir, Oliver Stone, E. O. Wilson, Paul and Anne Ehrlich, the founders and key thinkers behind Greenpeace, the World Wildlife Fund, the MacArthur Foundation's biodiversity grant division . . . these are just a few of the people and groups that inspire me. But I could cite thousands of animals I have come to know personally—these are true teachers of tolerance and compassion. Parrots like Mac, Feather, Stanley and dear Josie. I have written about some of these people in my book *Voices from the Underground: For the Love of Animals*.

Q9. RELIGION/SPIRITUALITY

I have referred to Jain tradition. But I would also suggest early Judaism, Taoism, and Christianity—certainly as interpreted by St. Francis and others—as great paths toward compassion. The spiritual traditions of the Bishnoi of Rajasthan—who worship a medieval environmental saint—and the 1200 tribal Todas of the Nilgiris in Tamil Nadu (south India) are also of utmost importance to understanding the possibilities of spiritual communities that are taking an active role in biological conservation and compassion. The Karens of Thailand and Myanmar; the Hadza of Tanzania; the Inner Badui of West Java, the Bhutanese . . . these are others of great importance to the world's natural heritage and our future.

Q10. FUNNY/STRANGE EXPERIENCES

Funny? No. Intolerably sad, every day.

Strange? To be a human. To recognize that—too frequently—I feel like an utter outcast, an alien, a nonmember of my species. Would I rather be a tree? Or bunny? On many days, yes. Would I forego the pleasures of Shakespeare and Bach and Mozart and Vermeer to become a cricket? On many days, yes. That is strange.

Q11. LESSONS/ADVICE

I don't believe there is anyone out there who does not love a beautiful sunrise or sunset. Or thrill to a good breath of air, a cool drink of water, a lovely sky, the onset of love. Even Hitler loved nature, as strange as it seems. I have looked at dozens of his landscape paintings, and there is no question he loved the world. It is also rumored he was something of a vegetarian (for "Aryan" reasons). My point is this: Even in the heart of evil there "may" dwell hope; what some would call hope for redemption, what others might better characterize as hope for persuasion. Gandhi misread Hitler in this way, however. He imagined that the Jews needed to simply lie down in passive protest. That the Nazis would soften and come to understand that the Jews were people, just like them. And that World War II would simply go away. George Orwell pointed out Gandhi's terrible simplicity on this matter. But in other ways Gandhi was astute, obviously. In the early 1900s, it was Winston Churchill who recommended independent status for India and all of the Commonwealth countries. He was ignored, though Gandhi certainly reminded the government in his day of Churchill's speeches and writing. We must study history, law, economics, politics, anthropology, art history, comparative literature, spiritual traditions and psychology in order to begin to recognize the cultural context in which any successful animal activism is likely to happen. Working in a cultural void is pointless. And the same goes for NGOs ("nongovernmental organizations") or NGI's (non-government individuals) working in any other country or community.

The lesson, then, is tolerance, compassion for other people, but a firm resolution in one's own heart. Never lose sight of what you're doing, where you're going; but grant that the path will be difficult and requires utmost patience. We must have courage to be patient and optimistic in these times. I am impatient by nature and so this has proven the most trying of all aspects of environmental activism. I rebel against patience every day, but—usually—it is not a helpful agitation. The key for me is trust— trust that what I'm trying to achieve is good. That I may not change the ocean's tides, but that my tears are meaningful, nonetheless. I make no effort to conceal them.

Q12. ADDITIONAL INFORMATION

Never give up. In the end, as in the beginning, love and compassion are key to life on Earth. Evolution does not condemn us. Only our choices will do that. Let us choose peace and nonviolence. Let us find the techniques to disseminate this message. Every moment provides a new opportunity to be loving.

Frankie L. Trull

Q1. BIOGRAPHICAL PROFILE

I am fifty years old (thereby ensuring I have absolutely no credibility with the high school students who may read your book).

I was born in Massachusetts; schooled in Massachusetts and Virginia. I graduated from Boston University, cum laude, History major and received a master's degree from Tufts University in Sociology. My first job was in the grants and contracts office at Tufts medical school in Boston.

Q2. BECOMING INVOLVED

In the mid-1970s I worked for the president of Tufts, and my primary responsibility was to assist in the establishment of a veterinary school, as New England was the only region of the country that did not have one. During the process of putting a curriculum together, the president of Tufts (Jean Mayer) wanted to put emphasis on areas of veterinary medicine where there was a perceived need, including nutrition, aquatic medicine, equine medicine, and laboratory animal medicine. During this time, I met a number of lab animal vets, and after the veterinary school was up and running, I was approached by some of the vets who were under attack by critics of laboratory animal research. This was 1979 and "animal rights" had not yet been coined. These were animal "welfare" activists and antivivisectionists. Because I had worked at an academic health center, I certainly understood the need for animal-based biomedical research, but I was also an animal lover—and have been a lifelong horse owner. So I made sure

Frankie L. Trull. Photo credit: Randy Santos, Randolph Photography

I had a good understanding of the animal research process—when, where, how, and so on—before I committed to setting up the National Association for Biomedical Research (NABR, 1979) and the Foundation for Biomedical Research (FBR, 1981).

I don't view myself as an activist, but rather an advocate. To me, there is no more noble partnership between man and animal than the partnering in research to better understand, treat, and cure diseases and injuries that afflict both humans and other animals.

Q3. IMPORTANT ISSUES

I spend most of my time doing legislative work and media interviews. I care about this issue because medical research is the underpinning of so many medical discoveries.

Q4. GOALS & STRATEGIES

Our goal is to ensure a climate conducive to allowing the pursuit of medical knowledge. This is accomplished through education, education, education. This takes place in many ways, through many vehicles, all of which we try to use. The fact is, this is a battle for the hearts and minds of the public. Without their support, much medical research would not proceed. But as long as there is sickness and disease, we will prevail. Humane and responsible animal research is a scientific and moral imperative.

Q5. DEFINING SUCCESS

[No response.]

Q6. GROUPS & ORGANIZATIONS

For me, one of the great benefits of representing NABR and FBR has been the opportunity to work with some of the great minds of our time. Dr. Michael DeBakey, the world-renowned heart surgeon, is chairman of the board of FBR. Additionally, a number of Nobel prize winners are also represented on our Board. They support our efforts because the fact remains that animal research is critical to medical progress. Even though animal rights activists deny this, they are totally wrong. But to help the public distinguish between scientific fact and emotion is a major challenge. And it is this challenge that I find most interesting. My responsibilities over the years have been and continue to include lobbying the U.S. Congress to ensure that legislative and regulatory proposals do not have an adverse effect on biomedical research, conducting public relations and public education programs, fund-raising, giving speeches, and doing interviews with the print and electronic media and general administrative duties for these two organizations.

Q7. "THE OPPOSITION"

I view all animal rights organizations who are opposed to animal research as the opposition. We learned long ago that there is no point in arguing with or attempting to convert them. They are entitled to their opinions. Our challenge is to reach the members of the public who are willing to listen and learn. And the public is far more thoughtful and responsible than they are often given credit for.

My organizations work closely with a number of groups who also care about these issues, from university organizations to disease groups. Broad coalitions have much more impact in conveying a message than do individuals or single organizations.

Q8. HEROES

I admire all the scientists who labor in laboratories searching for tiny clues which collectively add up to what we've all come to expect—medical miracles.

Q9. RELIGION/SPIRITUALITY

[No response.]

Q10. FUNNY/STRANGE EXPERIENCES

Because I've been representing this issue for twenty-plus years, there have been lots of interesting and/or strange experience. Driving into my parking garage and being attacked by activists was a very strange, and frankly frightening, experience. Winning a legislative victory is a high. Working with Congress on such an emotional and complex issue is very challenging.

Q11. LESSONS/ADVICE

My advice to anyone working on this issue, on either side, is to be factual. Attacking people or property is an immature, inappropriate way to get your message across. If it's not strong enough to stand on the facts, it's a losing battle.

Q12. ADDITIONAL INFORMATION

[No response.]

William L. Wade

Q1. BIOGRAPHICAL PROFILE

My name is William L. Wade, LVT, LATG. I was born in Jamestown, New York, on June 24, 1954; but grew up on the north shore of Long Island in Port Jefferson, a harbor town. After high school, I tried college at the SUNY at Alfred, but during the second year, I realized that I was not prepared. In February 1974, I enlisted in the U.S. Air Force as a veterinary specialist and served for ten years. In 1983 I transferred into the U.S. Army Veterinary Service and completed eleven years of active service. I retired in October 1994 having served over twenty years of active military service.

In November 1994, I accepted the position of manager of Veterinary Laboratory Animal Care at the School of Veterinary Medicine, at Purdue University. In March 1999, I accepted my current position as manager of Compliance & Training at Northwestern University in Chicago.

I enjoy spending time with my wife, taking short trips around the Midwest whenever we can, or visiting family in New York, Georgia, and South Carolina. Playing golf is my primary weakness, and I have been addicted to the game since childhood. We have also recently purchased my dream car and are in the process of restoring a 1969 Plymouth Roadrunner that we hope to take to auto shows throughout the Midwest.

Q2. BECOMING INVOLVED

My initial involvement came through service in the military at the Walter Reed Army Institute of Research. I was initially assigned as an instructor at the Animal Care Specialist School, the Army's equivalent of a program in Veterinary Technology. After four years in the school, I was laterally assigned as the NCOIC (manager) of the Department of Animal Medicine. Our primary role was to provide housing and veterinary care for a wide variety of animal species. I was an active and voting member of the Institutional Animal Care and Use Committee (IACUC), the appointed group that reviews and monitors research protocols and activity. This was my first immediate role within the arena of biomedical research involving animal models. It was my responsibility to ensure that our facilities were in compliance with the federal laws pertaining to the care and use of research laboratory animals. This included daily care and housing, veterinary medical care, and physical plant support capabilities. Our programs and facilities were inspected by both the USDA and the IACUC (as mandated by federal law), to ensure that we maintained compliance with all applicable standards. It became very evident that the welfare of laboratory animals was of paramount importance to the institution and that all measures needed to guarantee that they were kept and used in a humane manner were easily justified. You cannot feasibly maintain compliance in a facility by cutting corners.

I do not consider myself an animal "rights" activist; I believe the term is, at present, not defined that distinctly. By profession, I am obligated to ensure the health and welfare of all animals under my care. I am certainly an advocate of animal welfare. It is the hallmark of our profession and the primary responsibility that we have. In that respect then, I am always an activist in the sense that I must ensure that animal welfare is maintained regardless of the circumstances. To equate rights to animals in the sense that we have with human society is practically impossible. Animals by nature conform to an entirely different set of rules and behaviors. To maintain that animals can be afforded the same basic rights as humans does not compute. Freeing animals from research facilities will not go a long way in securing their future survival, certainly not as feral animals. My personal opinion is that there are far more pressing issues on the docket of animal welfare. Increasing pet overpopulation, indiscriminate breeding of animals for sale as personal pets (a.k.a. puppy mills), and illegal activities involving animals (e.g., fighting). These activities do nothing to promote the health and well-being of any species.

Q3. IMPORTANT ISSUES

I spend the great majority of my time (80 percent) verifying that everyone involved in the area of laboratory animal research is in compliance with the federal laws. This includes training, inspections, and consultations with all affected. This is, I believe, the most critical aspect within the workings of a research facility that uses animal models. The requirements are vast, and the ramifications of not being in compliance are less than desirable. The regulations and guidelines change consistently, often requiring that we retool systems and practices in order to conform with these changes. To the average person not directly involved with the use of research animals, these activities should be seen as sincere efforts to ensure that, as long as animal models are used for these purposes, their health and welfare is a top priority.

Q4. GOALS & STRATEGIES

Short-term goals are to continue to introduce viable training programs to all parties involved in this process. The long-term goal would be to expand these training initiatives to a broader audience (through Web-based programs for example).

In my role as compliance and training officer, I primarily do research, write, and present information on the humane care and use of animals in research settings, as well as provide guidance on assuring that all programs are in compliance. I also manage the programs in facility compliance to include physical plant parameters (room airflow, temp, humidity, and lighting), function of essential support equipment (sterilizers), and sanitation practices (through microbiological analysis after procedures). I work to develop programs and policies that allow personnel and facilities to maintain compliance with animal welfare laws.

Q5. DEFINING SUCCESS

I will sense success when no issues or incidents of noncompliance exist. This is difficult to assume given that the human component can never be extracted from the equation. It would be a significant accomplishment if we should ever see the day when the use of live animal models is eliminated by alternative applications or models. I suspect that given our technological capabilities someday this will occur. The advances to date in genetic and synthetic tissue culture activities is a positive sign.

Q6. GROUPS & ORGANIZATIONS

Yes, I work with several organizations, including NAVTA, American Association for Laboratory Animal Science, Office of Laboratory

Animal Welfare, American Association for the Accreditation of Laboratory Animal Care, and Applied Research Ethics National Association. I have served two terms as president of the North American Veterinary Technician Association (NAVTA). Most of these organizations are dedicated to the health and well-being of all animal species. Some in particular are regulatory agencies, which, through inspections and consultation, ensure that all institutions where animal research is conducted are in compliance with federal, state, and local laws.

Q7. "THE OPPOSITION"

Overzealous and extremist animal "rights" groups. Their focus is usually way off the mark. Their actions typically do not serve a long-term goal or purpose. I would like to see their efforts carried out in a more productive format.

Q8. HEROES

My parents above all. I've been inspired by several colleagues, especially those in the military who are too numerous to mention. In the field of veterinary medicine and technology, there have been many: Patrick Navarre, Carlene Decker, Dr. Jack Mara, and Dr. Rich Ford. All of them have had a major impact on my career and growth in this profession. I've found inspiration in the music of many artists, in particular the late John Denver, The Beatles, and Neil Young.

Q9. RELIGION/SPIRITUALITY

Religion is not a part of my work to any great extent. I have certain beliefs that allow me to associate with and appreciate all forms of life and in particular the creatures I provide care for.

Q10. FUNNY/STRANGE EXPERIENCES

I am not an activist, except to say that I am a proponent of welfare for all species.

My experiences have been more along the lines of disbelief in some of the things that so-called "educated" people do. I have found individuals in primate rooms not wearing protective equipment, when there was an imminent threat of exposure to fatal disease.

I have been present at several activist "demonstrations" and am constantly surprised and somewhat amused by their actions. I understand the point that they are trying to make. It just seems like locking yourself in a cage on a sidewalk, in the middle of summer, when it's 102 degrees Fahrenheit in the shade is not really getting the point across.

Q11. LESSONS/ADVICE

Never believe that you know all there is about a particular issue. Make sure that you have reliable information before taking a stance on anything. Most information being published about animal research is antiquated and exaggerated. Are there still problems? Certainly, but the key is to discover new ways to ensure that these practices are being carried out in a humane manner and that alternatives to the use of live animal models are being explored.

Q12. ADDITIONAL INFORMATION

As long as man is the supposed dominant species, animals will play a subservient role, whether as pets, food producers, research models, and so on. Destroying labs, farms, and out-of-control puppy mills, will not make this go away. Concentration of efforts on awareness of these issues will go much further than a single act of defiance. Releasing lab rats into the wild is a death sentence to them. Discovering alternatives to the use of rats for research eliminates the need to continue producing them. More federal support is needed to begin eliminating or regulating the true animal welfare issues of indiscriminate and unnecessary animal breeding, dog fighting, abused racing animals, and so forth.

Ed Walsh

Q1. BIOGRAPHICAL PROFILE

I was born the son of a mailman in Springfield, Illinois, where one hundred years earlier Abraham Lincoln practiced law, married, raised his children, and became the sixteenth president of the United States. In the winter of 1946, the year I was born, Springfield hadn't changed much since the time of Lincoln, or so it seems to me as I look back through the detachment of time. Aside from being the seat of state government, it was a slow-moving place where kids played sandlot football and baseball, and adolescent boys swooned over Annette Funicello. Everyone seemed to know everyone else in the Springfield of my youth. Folks there, at least those that I knew, were pretty down-to-earth, just more or less living to get by and trying not to be consumed by the ever-expanding sense of complexity in our culture.

Although Lincoln set civil rights on the stage of history, in the Springfield of my youth blacks were pretty much shoved off to the east side of town, segregated from us lower middle-class white aristocrats. All of this made me wonder just how much Lincoln really achieved in his search for uniform justice and universal human rights. Blacks were an uncommon sight on downtown Springfield streets in those days of my youth, and, to me, they seemed a caricature of all that Lincoln said they weren't. This sociopolitical contrast between the real and the ideal was a constant as I came of age, and it undoubtedly shaped me personally and formed the foundation of my attitude toward the idea of human rights. Over time, Abraham Lincoln became my personal hero, a man that I have tried most unsuccessfully to emulate in life. I narrate on this matter to make the point that while growing up in Springfield we spent a lot of time thinking about these weighty ideas, and in doing so came to understand the volatility of this uniquely human concept, and the extraordinary difference that one brush stroke of courage and creativity can make on and in our culture.

I'm a biologist who grew up in the 1950s and 1960s in the Midwest, an experience that may explain the wide range of my feelings, if not my bipolar ambivalence, on at least some social issues. After completing undergraduate studies in Biology and Chemistry, I taught high school Chemistry, Math, and Physiology at my alma mater during the period that the war raged in Vietnam. I was opposed to the war and was curiously never drafted even though I was highly eligible. Although I had never considered teaching as a career, and although formally (if not fully) unprepared for the job, I was invited to teach Chemistry, a position that opened unexpectedly following my graduation from college. Although hesitant, I accepted the offer (fully expecting to be drafted any moment) and was immediately hooked. The challenge was captivating and there was something unexpectedly satisfying in the process itself; it was clearly the most important thing I had ever done. The prospect of affecting the attitudes and dispositions of high school kids was at once humbling and inspiring. After a few weeks on the job, I could barely remember wanting to do anything else with my life. After completing a Master's Degree in the Biological Sciences, I accepted another teaching assignment at a small junior college where I taught Anatomy and Physiology, Comparative Vertebrate Anatomy, and a host of other more specialized courses to undergraduates and nursing students.

Although delighted being a teacher, almost from the start I was bothered by the fact that I was teaching Science and had never actually worked as a scientist. In those early days in the classroom, I came to believe, as I do today, that our plan for teaching Science, for teaching anything intended to prepare our children for life for that matter, is essentially flawed. We shouldn't be teaching teachers to teach. Rather, we should teach writers to teach Writing and Literature, economists to teach Economics, State Department operatives to teach History, and scientists to teach Science.

So, I quit my job and secured a position in the Graduate School at the Creighton University School of Medicine Department of Physiology and started working on a doctorate at the Boys Town National Research Hospital (then known simply as the Boys Town Institute), studying Neurophysiology. My goal was to gain real experience so that

I could transform the way Science was taught to high school students and undergraduates (to preschoolers for that matter) through example. I wanted to turn Science education around, placing emphasis on the philosophy and rationale of Science and teaching from the perspective of its practice, as opposed to the contemporary approach of teaching science facts exclusively.

With many regrets, I never got back to the classroom. The laboratory was for me magnetic, and I was drawn fully into research. Although I occasionally lecture in the medical school and devote much time to graduate students in my lab (a form of teaching—one on one—that I truly enjoy), teaching is secondary in my life and I miss it very much. However, the thrill of discovery is enchanting and the satisfaction that goes with adding even microscopically to knowing a thing is innately satisfying. Aside from the strain of publishing or perishing in one's field and the anxiety of living in a soft-money research culture, the life of a research scientist is wonderfully demanding. Aside from perhaps a seat on the Supreme Court, I can't think of a nobler, more important, and more rewarding job.

I enjoy writing poetry. Discovering new things. The feeling that what I do makes a difference. Quiet moments. Watching my son sleep at night—not that I don't enjoy him during the day. Watching JoAnn sleep at night—not that I don't enjoy her during the day. Walking in the woods. Photography, especially while walking in the woods or along a coastline somewhere. The feeling of energy that flows in the often turbulent interface between knowledge and ignorance. The beauty of high human achievement and high human drama. Inventions that are created when technology and nature and human daring intersect in time and space. The start of a day.

Generally, I dislike suffering of any kind. Personally, I dislike unfinished things, unresolved plans, unachievable dreams, vulnerability and inadequacy, not to mention making easily avoidable mistakes. The end of a day.

Q2. BECOMING INVOLVED

For me, at least, this is a pretty complex question. I can't remember a time that I wasn't interested in other animals. My father was a hunter and a fisherman, and my first recollections of other animals are tied to many early, brisk fall morning hunts for quail, rabbits, and pheasant, or blue gill fishing expeditions. Although it is popular in some quarters to claim otherwise, this experience gave me a sense of connection with the rest of the animal kingdom. I came to understand our biology, and respect it, more in those experiences than in any Biology course that I have taken since. The hunters and fishers that I knew were deeply compassionate men (I didn't know any women hunters, although I knew of them), and the act of hunting and fishing was much more about bringing balance and meaning to biology than it was about killing animals. For whatever reason, I came away from my hunting experience a nonhunter with an abiding respect for hunters and for animals and the natural world generally.

Although I knew of the animal rights movement for many years before being attacked by it, like most of my colleagues I was just too busy to worry much about it. I fooled myself into believing that I was safe because I treated my animals humanely and respectfully. All of that changed dramatically after I became a target of the movement.

Even now I don't think of myself as an activist. I am a scientist who was hit between the eyes with the animal rights "two by four." I was shaken awake and forced to recognize that a social movement was threatening to reshape our culture and the way that we conceive our biology. Fully unprepared to articulate the philosophical arguments necessary to counter the animal rights rhetoric and their blatantly specious arguments, I nervously jumped into the fray. Now, I nervously stay with the debate—nervous that I will fail to be clear and effective in the articulation of my argument.

Q3. IMPORTANT ISSUES

I am still recovering from a remarkably successful animal rights campaign against us. I spend some of my time pointing out the absurdity and the danger of lawyers, political scientists, and philosophers with a sociopolitical agenda predicated on unfounded biological principles. I worry a lot about the leaders of a movement who are either inadequately prepared to comment on issues of our biology or who are frightfully disrespectful of it in their political ambitions. I also worry a lot about sensitive, well-intentioned humanists who just don't seem to understand the difference between animal rights as a philosophical and political tool, and animal welfare.

Dishonesty infuriates me, and my fear that the tyranny of the minority can be achieved in a lethargic democracy reaches a peak when I think about the animal rights movement—undoubtedly the result of my personal experience with their social and political agenda. I believe very strongly in pluralism, but fear the potential of factions when political will is unbalanced.

The quality and meaning of their lives hang in the balance.

Q4. GOALS & STRATEGIES

My short-term goal is honest debate. My long-term goal is honest debate. Maybe the advocates of animal rights are correct. How will we ever know if they continue avoiding civil, objective and factual dialogue? I claim that the hard-core leaders of the movement avoid honest, scholarly debate because they recognize the essential flaw, the essential dishonesty of their movement, that animal rights is not about animals, it is about political power. Unless animal rights advocates can explain on rational grounds how an amoeba or a paramecium is the biological/ethical equivalent of your son or daughter, their philosophical argument dissolves; their effort to reformulate biological principles, which is an essential element of their model, is discredited.

I write a little. I talk about the issue with as many people as possible, always trying to be fair to both points of view by explaining the essential differences, as I see them, and my reason(s) for accepting animal welfare over animal rights. Having been the recipient of the wrath of public protest, my enthusiasm for its utility as a social tool has waned. I have never broken a law, I care about the animals that I use, and I respect diversity of opinion as the essential driving force of social evolution. In spite of all this, my son had to endure protests at our home, threatening phone calls and letters, and a general strangeness in living during a critical stage of his development because a political faction disagreed with the choices that his father and mother made in life. All of this has diminished my enthusiasm for social protests aimed at law-abiding individuals. Although I do not advocate placing limits on free speech, there is clearly a difference between the right to speak freely about someone and harassment. That difference can be subtle and may require legal clarification as we grapple with the question of rights in an increasingly complex culture, but we must deal with it. Nonetheless, I quite literally long for the day that people with disparate views learn to tolerate and respect the differences in our culture. That would be my version of heaven. On this point, however, I am not as naive as I sound, and I'm certainly not holding my breath, anticipating such a profound transformation in human politics. But let me be clear on one very important point. I would not change my life plan in any way to accommodate any faction, nor will I stop studying animals to advance the human cause to avoid the discomfort that the animal rights community has heaped on my family.

Picketers around our home, death threats, personal and intentional misrepresentation, and sleepless nights are small prices to pay for an open, free society. I do, however, think that we need to work a little harder on the right to privacy clause of the human contract.

Q5. DEFINING SUCCESS

I will see success when there is honest dialogue and respectful participants on both sides of the argument. Knowing the attitudes and personalities of the leaders of the animal rights movement, it is almost impossible to imagine this happening; however, we can find some encouragement in the fact that even Arafat is now talking to Barak. My short term, more realistic goal is to save the animal rights community from themselves, along with the rest of us, by blocking their effort to bring biomedical research requiring the use of animals to its knees and the destruction of animal enterprise around the world.

Q6. GROUPS & ORGANIZATIONS

I am fortunate to serve on the Board of the National Animal Interest Alliance (NAIA). I have great respect for, and am forever indebted to, the National Association for Biomedical Research and the Americans for Medical Progress for their support during those confusing and frightening weeks and months following PETA's targeting of me, my wife, my children, and my way of life. They know (I hope) that they can count on me to do what I can to support their efforts. I am convinced that NAIA will step in to fill the national vacancy left when the Humane Society of the United States abrogated its responsibility as an animal welfare leader and stepped squarely into the animal rights camp.

Q7. "THE OPPOSITION"

My opponents are any one, or any group, that chooses sensationalism, disrespect for biology, and dishonesty in their effort to achieve goals.

Q8. HEROES

Edward O. Wilson, the father of sociobiology and, in my opinion, the finest thinker of our culture.

Pythagoras. Driven toward enlightenment and consilience (to borrow a wonderful term that is the title of Edward Wilson's extraordinary book), his philosophy of living offers the only hopeful path in human culture. Abraham Lincoln's courage to advance the ethereal cause of human rights inspires me, and Gandhi's approach in overcoming oppression gives me hope in and for humankind.

Q9. RELIGION/SPIRITUALITY

I am spiritual, even in my rejection of spiritualism. Our human capacity to consider issues beyond ourselves is the source of our selective advantage, and is probably the selection pressure that will be our undoing.

I am not really "religious." I was raised a Roman Catholic and being reared in that culture my character was obviously shaped, in part, by the experience. However, having been sensitized to the Judeo-Christian mode of thinking, I try very hard to think outside of that box.

Q10. FUNNY/STRANGE EXPERIENCES

Absolutely none.

Q11. LESSONS/ADVICE

Be honest.

Be a realist. Don't believe for a heartbeat that you are immune from treachery because you choose anonymity and honesty and compassion in your life. The forces of human evil are genomic, and the drive to acquire power and influence are blind. In this, the truth seekers of our culture suffer a disadvantage in our biology.

Q12. ADDITIONAL INFORMATION

When you think about the plight of animals, begin by thinking about the plight of your children. Think about the plight of women dominated by men, the plight of cultures dominated by other cultures, and the plight of the nonreligious at the hands of the religious. Care for the animals, don't shun your brother and sister, and, most important, study biology—and then think.

Ben White

Q1. BIOGRAPHICAL PROFILE

I was born into a southern military family in Portsmouth, Virginia, in 1951. Both sides of my family come from the tidewater area of southern Virginia, right near the border with North Carolina. My mom's dad was a longtime guide for hunters and fishermen in Back Bay, just south of Chesapeake Bay; her mother taught piano and kept house for the seasonal hunters. My other grandfather was a boatbuilder, and my grandmother ran a rooming house for young Navy couples in Norfolk, Virginia. My roots are rural and maritime, but because my dad was a career Air Force officer, I lived all over the place as a kid, mostly in the Midwest and Spain before my family moved to the Washington, D.C., area in the early sixties. I think that some of my allegiance to nature comes from always being the new kid on the block, seeking companionship with critters and wild places instead of relying on long-lasting human friendships.

In the winter of 1969, I quit the University of Virginia after just a semester of "higher learning" to learn to be a "tree doctor," now known as an arborist. I climbed trees for a living for over twenty-five years, setting up my own company in Northern Virginia. The company specialized in saving trees affected by construction damage. It was quite successful: pioneering new ways to rescue root damaged trees. After running the company for eighteen years, I sold it to my employees, and dedicated myself full time to animal protection. Well, not quite full time. In 1991 I separated from my wife and obtained sole custody of my two children. They have lived with me ever since. My daughter is now fourteen and my son eleven. They are the lights of my life.

I enjoy lots of things: watching videos with my kids (we don't have television), climbing trees (still), listening to music (from Beethoven to Bob Marley), walking through the woods, bird-watching, swimming in the ocean, working with wood, traveling, writing, reading, debating, public speaking.

Q2. BECOMING INVOLVED

Personally, I don't like or use the term "animal rights" because it smacks of a human-centered ideology, as if we are "awarding" rights to the rest of the living world out of noblesse oblige.

For some reason, even though I came from a background of people who routinely used animals for food, I always had a soft spot for suffering creatures, whether children, homeless folks, or dogs and cats. Once my dad stopped someone from beating "their" dog. He is not by any stretch an animal rights activist, but his stand to prevent suffering stuck with me. My mom also gave me a profound and enduring adoration of nature, especially trees and flowers.

A turning point came for me in 1971 when I swam alone with a pod of wild spinner dolphins off the coast of Hawaii. A penetrating gaze from one of the dolphins crashed my existing worldview, which held all creatures as essentially interesting stuff. This view was replaced by the perception that dolphins and all creatures are self-aware conscious entities like myself, even though encased in quite different bodies with very different minds. This led to my conviction that people are only a lonely fragment of the Earth's greater family of life. Instead of treating sentient beings as stuff to be hurt and sold as we wished, our own path toward belonging lay in embracing this family as our own. On that day, I swore to the dolphins that I would use my life to protect them from harm from my own species. That vow, also extended to other marine mammals and terrestrial creatures, has been the essential motivator of all of my animal protection work ever since.

Q3. IMPORTANT ISSUES

Tough to boil it down to one or two. Most of my time is dedicated toward the global protection of whales and dolphins. I am focusing primarily on the two issues I believe are the greatest threats to these creatures: the development and deployment of low-frequency-active sonar (LFAS) by the U.S. Navy (and others) and the adoption of a

Revised Management Scheme (RMS) by the International Whaling Commission, which will legitimize the resurrection of full-blown commercial whaling. I also deal with old-growth forest protection and am involved in the international populist movement opposing the increasing corporate control of our lives. This grew directly out of my involvement with the demonstrations against the World Trade Organization in Seattle in November 1999. I was the creator of the herd of 240 people dressed as sea turtles that swamped the streets in protest of the WTO's theft of the U.S. law protecting sea turtles.

I care for the same reason that anyone would care if bullies were terrorizing their neighbors. My frame of reference is just larger than the norm. Every day I am flooded with faxes and e-mails telling of new crises affecting marine mammals. It is impossible for me to just ignore the suffering if there is anything I can do to stop it.

Perhaps oddly I do not see my job as an activist as telling people what they should or shouldn't do. I see my job as personally responding to stop suffering as best I can, hoping that by example I motivate others to become peaceful warriors for the greater Earth community. I don't think that preaching or guilt-tripping works. People can only respond when something touches them personally. Your gut either resonates or it doesn't. Generally, animal activists tend to be empaths—people who willingly take on the pain of others. Unfortunately, the world pain is too great and most eventually burn out.

Q4. GOALS & STRATEGIES

My attitude toward my work is summed up in a quote from Wendell Barry (in the book of essays called *What Are People For?*). He said (to paraphrase) that the true purpose of protest that succeeds is not the changing of all minds but to hold onto that within our heart that dies through acquiescence. I do what I do to refuse to acquiesce to dying the death of a thousand little personal compromises. My commitment comes from a promise between myself and the natural world. Therefore, whether I am able to rally hundreds or thousands to my cause, and whether my work is picked up by the mainstream media is essentially irrelevant. I do it for me.

My work is a curious straddle between the mainstream and the underground, between the legal challenge of government regulators and the often illegal challenge of bad laws and policies that hurt animals. My principle orientation is direct action, usually involving the insertion of my body between the violence and the victim. It amazes me that a person attached to a tree slated for destruction is deemed more newsworthy than a Goshawk or a Murrelet, but that appears to be the case.

For almost every crisis that comes along, the recipe for stopping the harm is the same. First is extensive research. Who is doing the harm? What laws apply? Who are the players on each side? What are the government regulators doing (or more likely ... not doing?). What is the precise point of injury? I believe that the power of a protest is directly proportional to the proximity to the injury being done. For example, directly stopping the clubbing of a seal, or the harpooning of a whale (especially when filmed and distributed) delivers a much more understandable message than a hunger strike in front of the Canadian or Japanese embassies.

Once I have crammed all the information I can find about the subject, I usually travel to the site of the problem, organize local folks, and try to directly stop the injury. I then write articles, contact Congresspeople, their staffs, and regulators to spread the word and clarify the argument. This often goes on for many years. So, for me, the answer is yes: research, educate, protest, and write; often in that order.

Q5. DEFINING SUCCESS

With the Navy's LFAS (Low-Frequency-Active Sonar), success would come with the Navy deciding to cancel the development and deployment of this unnecessary and awesomely foolish device which threatens all ocean creatures with some of the most intense sounds ever created by humankind.

With the RMS (Revised Management Scheme), success would see the abandonment of trying to legitimize commercial whaling in favor of the far more lucrative whale watching. The International Whaling Commission should evolve into a conservation body that protects both small and large cetaceans (whales and dolphins), and should permanently embrace a ban on all whale killing except that necessary for true aboriginal subsistence.

With the "antiglobalization" movement, success would involve the strengthening of grassroots democracy around the world; the dethroning of corporate power as embodied by the World Trade Organization, the International Monetary Fund, and the World Bank; and the creation of labor/consumer alliances that protect the Earth, protect national sovereignty, protect workers and local cultures, and protect our fellow creatures. Yes, quite a mouthful!

Q6. GROUPS & ORGANIZATIONS

I am international coordinator of the Animal Welfare Institute, based in Washington, D.C. On behalf of AWI, I attend meetings of the International Whaling Commission and the Convention on International Trade in Endangered Species (CITES). In that context, and in my frequent campaigns, I work with a very wide variety of groups including the Humane Society of the United States, People for the Ethical Treatment of Animals, Animal Legal Defense Fund, Earth First!, the Cetacean Freedom Network, and many, many others.

Q7. "THE OPPOSITION"

My work entails building arguments against the ideologies and behaviors of individuals and groups, as distinct from opposing them as bad people. I try to separate the two, believing that all of us are badly fumbling mortals and everyone is a potential convert. Oddly, I have found that I can get along just fine with whalers and those who kill for a living. Our disagreement is clear and honest. I have far less affinity with colleagues who I feel are betraying their cause and their supporters.

My principle opponent is a mind-set rather than the individuals that happen to be infected with it. Part of that mind-set is the belief once called "wise use" and now repackaged as "sustainable use." This philosophy holds that human beings are just another part of nature and that anything we do is "natural," even if it causes widespread damage or the extinction of wildlife. Many hold this philosophy and work to oppose any effort made to enact permanent animal protection. Wildlife is considered just another "resource." Believers include many corporate giants such as Weyerhauser Lumber Company and Anheuser Busch (owner of Sea World) and phony wildlife protection groups such as the World Wildlife Fund.

Q8. HEROES

Can't do just one or two. From ancient to modern: Lao Tzu, St. Francis, Joan of Arc, William Wallace, Gandhi, Aung San Suu Kyi, Tich Nhat Han, Ralph Nader, Julia Butterfly, Rod Coronado. Those people (mainly women) who dedicate their entire lives to keeping their families together—unsung and unnoticed.

Q9. RELIGION/SPIRITUALITY

I consider spirituality, which I would define as a love affair with life itself, as the central axis of my life and all of my work. I do what I do for love, which I consider identical to the creative force.

About five years ago, I participated in an Earth First! attempt to stop the cutting of a wildlife refuge called Rocky Brook, on the Olympic Peninsula. Only fifty-five acres, it had been set aside forever but was then given as a plum to timber interests as part of Clinton's "Salvage Timber Rider." I was arrested three times violating a "timber closure order." This closed a huge area around the clear-cut, supposedly for safety but actually to keep protestors and press away.

Finally I was strapped in manacles and chains and put in a federal prison for violating that law. In challenging my arrest and detention, we argued that my constitutional rights to speak, assemble, and worship in a place I considered sacred were being illegally infringed upon. The judge agreed, I was acquitted along with the over two hundred people arrested, the forest closure law as interpreted was struck down as overly broad and capricious. This gave me the idea that religion was a perfect vehicle for protecting wild places and creatures, especially in the United States where the concept of freedom of religion was originally founded. So I incorporated (in the state of Washington) the Church of the Earth to use the direct action defense of the Earth and her greater family as our sacrament.

I believe that our greatest human crisis right now is our spiritual disease that separates us from all life. This is the cause of our epidemic of loneliness and disconnection. We are caught up in a continual agitation to fill the void within our hearts with stuff, or partners, or TV, or computers because we feel so hollow. None of it works, and our voracious appetite is eating up the natural world piece by piece.

To me, the whole living world is sacred and life itself is the ultimate mystery and miracle. We are the generation on watch while it is all put on the auction block and sold forever. Our great failing is that we apparently just don't love this place enough to save it from our own hunger and confusion. My own personal spirituality involves loving this place, and my children, and the wild ones enough to risk dying for them.

Q10. FUNNY/STRANGE EXPERIENCES

Oh, I've got a million of those stories. Here's a funny one. For years, we had been protesting the captivity of dolphins and beluga whales at the Baltimore Aquarium. We were always good little boys and girls and kept to the area the police said we could use—far across a four-lane highway from the aquarium—where our signs and presence could be easily ignored. One beautiful spring day, I saw hundreds of people waiting in line for their tickets at the aquarium across the street and rallied up the protesters to take the daring move to actually cross the street so we could leaflet and speak to the people who were financing the captivity.

We trooped across the street. I stood up on a trash can like a labor organizer and started giving my rap about the cruelty of holding whales and dolphins away from freedom and family. After about ten minutes or so, a bored Baltimore cop sauntered over and said, "What do you think you are doing? This is a public park, there is no public speaking allowed here." So, for the time being I complied.

After pulling a copy of the ordinance he was referring to, I decided that it was a good one to challenge. The park ordinance prohibited a whole list of activities unless a permit was in hand, including "one or more persons sharing views." Of course, permits were not issued for aquarium protests.

So I returned one day with a friend named Logan Cockey to deliberately challenge this law. Logan was dressed in a striped prison suit with a dolphin head. He locked himself by the neck to the entry turnstile as I stood next to him bla-bla-bla-ing about captivity (this becomes hard to do after about thirty minutes). The aquarium personnel gathered. The police gathered. They asked me to stop. I kept talking. They asked Logan to release himself. He, dolphin-like, didn't respond.

Finally they came with a paddy wagon, cut Logan off his turnstile, and handcuffed the both of us. As they are walking us to the paddy wagon, one of the cops in charge said to the officers holding Logan, "Keep a good hold on that guy with the dolphin head, I wouldn't have to put out an all-points-bulletin with his description."

Here is a strange (and wonderful) story. Once I broke into the dolphin facility at the Hawk's Cay Marina in the Florida Keys. In the middle of the night, I cut open the chain link fences that divided the facility to access the main dolphin pool where eight or ten dolphins had been long kept. My destination was a newly caught dolphin being held in an inner pen. To get to him, I had to cut three fences and swim across the main dolphin pool. The phosphorescence was so bright that night that every clip of my one-handed bolt cutters through the chain link caused a brilliant green flash as the plankton was disturbed. I cut a hole the size of a Cadillac in each fence as I worked my way toward Hekili.

He had been caught about three months earlier in an illegal capture by Jay Sweeney. The dolphin caught with him had died after ninety days of forced feeding through a tube. The one who died was named, sardonically, Benny (after me). I was determined to give Hekili at least a chance to swim free.

After all of the fencing had been opened over a couple of hours of work, I was on my way out when I had to pass again through the main dolphin enclosure. For some reason, the dolphins mobbed me, with one coming up against each knee and elbow, almost pushing me out of the water. I have never seen this behavior before or since. I have no idea what they were saying to me. As I looked down into the pitch black water I could see dolphins swimming under me, their entire bodies glowing green with phosphorescence. It was a sublime and eerie sight I shall never forget.

Q11. LESSONS/ADVICE

As a young person I was bullheaded, arrogant, and individualistic and never accepted any of the good advice given me. I always insisted on doing things the hard way. I see no reason why any beginning activist should listen to pearls of wisdom from me.

That said, of course I will throw a few out, just for fun:

Take time to be nourished by the real outside, instead of the abstract, even if it is just a weedy lot behind a shopping center.

Don't take yourself too seriously. Don't abandon humor, music, and fun. If you are too deadly serious, potential converts will run.

It ain't about you. No matter how many stories show your pretty face. If you think it is about you, you will distract rather than lead.

Be easy on your friends and family. You can't force anyone to care. We all have different ways.

Be easy on fellow activists if they don't choose your path. Our diversity of approach is our strength. Avoid competitive purism. Unless you grow all of your own food and walk everywhere, you too are contributing to the destruction of the wild. Don't expect to be perfect. Our work is a path, not a destination.

Set incremental attainable goals. Educate yourself endlessly.

Don't exaggerate. The truth is bad enough.

Remember death over your left shoulder. Live life as if it is your last day. Live life so as to have no regrets.

For years I did my opponents the favor of seeing myself as a radical. That came, I suppose, from my high school days when, like many,

I was trying to figure out who I was. Part of the conceit was that I was better than others because I was taking a stand. That's a lot of hooey. I now have come to see that I am really a conservative—I wish to conserve what we have. My interests are protecting home and family, no different than anyone else. If I had realized this earlier I might well have been more persuasive.

I no longer believe that carrying signs and angrily chanting works to convince anyone of the justice of our cause. I much prefer songs and flags.

Q12. ADDITIONAL INFORMATION

Our greater family is calling on us to join in the big dance. That is the way toward healing our spiritual disease. As long as we see ourselves as the crown of creation, and all other life on Earth just stuff (or "resource"), we will be lonely and afraid.

We are in the end game. With over 6 billion very hungry and aggressive human primates on Earth, our materialism will devour everything that is wild and wonderful if we don't value it and protect it with our lives. It's your choice. Would you rather your children and grandchildren live in a forest or a factory? Every single action and thought have ripples that spread forever.

This is not a rehearsal. This is the real thing, and we won't get a second chance.

Appendix A: Interview Questions

Q1. Could you tell us a little bit about yourself? Who are you? Where and when were you born? What do you enjoy? [BIOGRAPHICAL PROFILE]

Q2. How did you become involved in animal rights issues? Was it a single event, or a gradual process, that started you down the path toward activism? [BECOMING INVOLVED]

Q3. What are one or two issues that you spend the most time on? Why do you care so much about this (or these) issues? Why should other people also care about this? [IMPORTANT ISSUES]

Q4. What are your short-term and long-term strategies for achieving your goals? What do you do exactly, on a regular basis, toward these goals? Write? Research? Educate? Protest? [GOALS & STRATEGIES]

Q5. What is your ultimate goal on this issue? When could you say, "we have succeeded?" [DEFINING SUCCESS]

Q6. Do you work closely with any formal groups or organizations? Which one(s)? [GROUPS & ORGANIZATIONS]

Q7. What groups or types of people do you consider to be "the opposition?" ["THE OPPOSITION"]

Q8. Whom do you admire the most, or who inspires you? Name one or two, modern or ancient heroes you have. [HEROES]

Q9. Is religion or spirituality a part of your life? Does this religion (or lack of it) help, motivate, or hinder your work? [RELIGION/SPIRITUALITY]

Q10. Would you describe a very funny or very strange experience that you have had in your work as an activist? [FUNNY/STRANGE EXPERIENCE]

Q11. What advice would you give to a person just starting into animal rights (or opposing-animal-rights) activism? What important lessons have you learned; or mistakes have you made, that others might learn from? [LESSONS/ADVICE]

Q12. Do you have anything to add to these questions that might be helpful to readers? [ADDITIONAL INFORMATION]

Appendix B: Participant Letter

Dear Prospective Participant,

I am John M. Kistler, professional librarian, and writer. I am under contract with Greenwood Publishing Group (Westport, CT.), working with editor Emily Birch, to write three books on animal rights issues. The first book, *Animal Rights: a Subject Guide, Bibliography, and Internet Companion*, was published in June of 2000. The Foreword was written by Marc Bekoff.

The second book is tentatively titled *People Promoting & People Opposing Animal Rights: People Making a Difference*. This will be part of a series of books called "People Making a Difference," all of which will be comprised of interviews with activists from different fields. Obviously, my volume will be comprised of interviews with animal rights activists and their opponents. The target audience will be high school and adult readers. The purpose is not to exhaustively discuss all of the animal rights issues, but to show "the human side" of activists; that activists are "real people" who are not all that different from the rest of "us." Rather than arguing the philosophies or merits of the animal rights issues, we want to reveal what the involved characters are like. We often view activists and issues wholly in the abstract; they have no identities aside from their role in the political games. My goal is to de-abstract the players; to show that they are normal folks who have taken a larger-than-normal role in an issue, and how they got to be there.

In case you are wondering, I am here defining an activist as a person who has taken a greater-than-average role in promoting or opposing an animal-rights-related issue. The media sometimes portrays activists as radicals. By my definition, a radical may be an activist, but an activist is not necessarily a radical. I consider myself to be an activist, because I spend much of my time considering and writing about animal-rights issues.

I am writing you to ask if you would be available and interested in being an "interviewee" in this project. I may have already approached you by electronic mail regarding this project; this letter then is a more formal request, with more relevant details for your consideration. I am seeking to enlist interviews from a variety of sources across the whole spectrum of animal rights issues (and their opposition). I have noticed something either in your personal history or the issues you are interested in, that encourage me to request your participation. [It is possible that some unforeseen issue might force me (or the publisher) to withdraw your participation from the project, though I doubt such would occur. Such a thing might happen if, for instance, you appear on the evening news performing vivisections, when your essay represents you as an anti-vivisectionist. I simply add this sort of stipulation because I wish to protect the integrity of this project, and leave myself free to cancel any essays which would jeopardize its purpose.]

I would guess, based upon the specifications from the publisher, that I will interview about 50 people, and that each person's interview will contain about 2,500 words. We would also like to include one black and white photograph with each chapter, when possible.

I was able to complete the first book three months earlier than planned, and thus I am getting an early start on the second book. For this reason, there is a larger-than-normal time frame to work with; and the benefit to you is that there is no major hurry. This book will not require personal face-to-face interviews. My plan is, when possible, to do the interviews chiefly by electronic mail or "snail-mail," with phone calls if necessary. The attached list of questions [Appendix A] will be sent to each interviewee, by e-mail or regular mail, and can be responded to in any way desired. I would hope that you might be able to complete your response in one or two months. Then I will edit the responses, and perhaps ask you for some clarifications. If you request it, I could send you back my edited piece (of your work) so you could feel comfortable with its content. This work will not be created or intended to be used in a partisan manner. There is no "right" and "wrong" belief in this book (unless you make glaring factual errors… like setting the world's population at 3 million rather than 6 billion). The point is to enlighten readers as to the people behind the issues, not to sway their opinions. That does not mean that participating people cannot express their persuasive ideas; it only means that my editing will not be used to slant those expressions. My editing would be chiefly for reasons of space, reduction of redundancies, and excluding material that does not seem relevant to the goals of the book. If you are not comfortable with any of the questions, you may skip them. Of course, if you skip too many, I may not be able to use your interview to meet the goals of the project.

The final question is intentionally very broad, enabling you to add anything of importance that I may have missed in the prior queries.

There may be a couple of interviewees who will remain anonymous or use pseudonyms, due to the nature of their activism. Their interviews do not imply any complicity on your part or my part with their activities. I will make

Appendix B

a strong statement to this effect in the Introduction. Our purpose is to share varying perspectives and experiences, even when they differ from "the norm."

Please respond to me quickly with your intention to participate or not to participate, so that I can find alternate interviewees as necessary. If I have not heard from you within one month, I will assume that you either did not receive my mailing, or do not wish to participate.

One last item that we need will be a black and white photograph of you, to put above the interview.

I believe that this will be a good opportunity for you, an activist (pro or con), to share some important lessons with the next generation of activists. Thank you very much for your consideration, and I look forward to hearing from you.

Sincerely,

John M. Kistler

[Editor's Note: this was the basic form of the letter. Some prospective participants asked for further information that was inserted into this letter or added in postscript.]

Appendix C: Organizational Addresses

Here are the addresses of most of the organizations mentioned by the contributors. Many other pro-animal-rights organizations can be found at the excellent website "World Animal Net Directory" at [*www.animalnet.org*]. Many other anti-animal-rights organizations can be found linked from the National Animal Interest Alliance website at [*www.naiaonline.org*].

Alliance for America
P.O. Box 449
Caroga Lake, NY. 12032
www.allianceforamerica.org

American Humane Association
63 Inverness Drive East
Englewood, CA. 80112-5117
www.americanhumane.org

American Society for the Prevention of Cruelty to Animals
424 Easy 92nd St.
New York, NY. 10128
www.aspca.org

Americans for Medical Progress
908 King St., Suite 201
Alexandria, VA. 22314-3121
www.ampef.org

Animal Liberation Front Press Office
BCM Box 4400
London, WC1N 3XX
United Kingdom
www.animalliberation.net

Animal Protection Institute
2831 Fruitridge Rd.
Sacramento, CA. 95820
www.api4animals.org

Animal Rights Advocates of Upstate New York
P.O. Box 18415
Rochester, NY. 14618

Animal Rights Legislative Action Network
www.animalpolitics.com

Animal Rights Online
P.O. Box 7053
Tampa, FL. 33763-7053
www.geocities.com/RainForest/1395

Animal Welfare Institute
P.O. Box 3650
Washington, D.C. 20007-0150
www.awionline.org

Animal's Agenda Magazine
P.O. Box 25881
Baltimore, MD. 21224
www.animalsagenda.org

Animals Voice Online
420 East South Temple #240
Salt Lake City, UT. 84111
www.animalsvoice.com

Ark Trust
P.O. Box 8191
Universal City, CA. 91618-8191
www.arktrust.org

Association of Veterinarians for Animal Rights
P.O. Box 208
Davis, CA. 95617-0208
www.avar.org

Boys Town National Research Hospital
555 North 30th St.
Omaha, NE. 68131
www.boystown.org/btnrh

Bushmeat Project
P.O. Box 488
Hermosa Beach, CA. 90254
bushmeat.net

Appendix C

Center for the Defense of Free Enterprise
Liberty Park
12500 NE Tenth Place
Bellevue, WA. 98005
www.eskimo.com/~rarnold/index.html

Coalition to Abolish the Fur Trade
P.O. Box 822411
Dallas, TX. 75382
www.banfur.com

Collective Humane Action and Information Network
P.O. Box 5651
Novato, CA. 94948-5651

Compassion for Animals
P.O. Box 72064
Corpus Christi, TX. 78472-2064

Dairy Education Board
325 Sylvan Ave.
Englewood Cliffs, NJ. 07632
www.notmilk.com

Dolphin Alliance
P.O. Box 510273
Melbourne Beach, FL. 32951
www.enviroweb.org/ahimsa/tda

Doris Day Animal League
227 Massachusetts Ave. NE, Suite 100
Washington D.C. 20002
www.ddal.org

Farm Animal Reform Movement
P.O. Box 30654
Bethesda, MD. 20824
www.farmusa.org

Farm Sanctuary
P.O. Box 150
Watkins Glen, NY. 14891
www.farmsanctuary.org

Federation of American Scientists for Experimental Biology
9650 Rockville Pike
Bethesda, MD. 20814-3998
www.faseb.org

Feminists for Animal Rights
P.O. Box 41355
Tucson, AZ. 85717-1355
www.farinc.com

Fishermen's Coalition
826 Orange Ave. #504
Coronado, CA. 92118

Friends of Animals
777 Post Road
Darien, CT. 06820
www.friendsofanimals.org

Fund for Animals
200 West 57th St. 10019
New York, NY. 14891-0150
www.fund.org

Fund for the Replacement of Animals in Medical Experiments
Russell & Burch House
96-98 North Sherwood St.
Nottingham, NG1 4EE
England
www.frame-uk.demon.co.uk

Fur Commission USA
PMB 506, 826 Orange Ave.
Coronado, CA. 92118-2698
www.furcommission.com

Fur Council of Canada
1435 rue St. Alexandre, Suite 1270
Montreal, Canada, H3A 2G4

Greenpeace USA
702 H Street NW.
Washington, D.C., 20001
www.greenpeaceusa.org

Humane Farming Association
P.O. Box 3577
San Rafael, CA. 94912
www.hfa.org

Humane Society of the United States
2100 L Street NW
Washington, D.C. 20037
www.hsus.org

Appendix C

Iditarod Trail Committee
P.O. Box 870800
Wasilla, AK. 99687
www.iditarod.com

In Defense of Animals
131 Camino Alto, Suite E
Mill Valley, CA. 94941
www.idausa.org

India Project for Animals and Nature
4912 Sherier Place NW
Washington, D.C. 20016

International Fund for Animal Welfare
P.O. Box 193
411 Main St.
Yarmouth Port, MA. 02675
www.ifaw.org

International Union for the Conservation of Nature
Rue Mauverney 28
1196 Gland
Switzerland
www.iucn.org

International Whaling Commission
The Red House
135 Station Rd
Impington, Cambridge
United Kingdom CB4 9NP
www.iwcoffice.org

Inuit Circumpolar Conference
www.inusiaat.com

Jews for Animal Rights
255 Humphrey St.
Marblehead, MA. 01945
www.enviroweb.org/jar

Last Chance for Animals
8033 Sunset Blvd. #35
Los Angeles, CA. 90046
www.lcanimal.org

Medical Research Modernization Committee
3200 Morley Rd.
Shaker Heights, OH. 44122
www.mrmcmed.org

National Animal Interest Alliance
P.O. Box 66579
Portland, OR. 97290-6579
www.naiaonline.org

National Anti-Vivisection Society
53 W. Jackson Suite 1552
Chicago, IL. 60604
www.navs.org

National Association for Biomedical Research
818 Connecticut Ave. NW, Suite 200
Washington, D.C. 20006
www.nabr.org

North Carolina Fisheries Association
P.O. Box 12303
New Bern, NC. 28561
www.ncfish.org

Orange County People for Animals
P.O. Box 14187
Irvine, CA. 92623-4187
www.ocpa.net

People for the Ethical Treatment of Animals
501 Front St.
Norfolk, VA. 23510
www.peta-online.org

Physicians Committee for Responsible Medicine
5100 Wisconsin Ave., Suite 404
Washington, D.C. 20016
www.pcrm.org

Professional Rodeo Cowboys Association
101 ProRodeo Drive
Colorado Springs, CO. 80919
www.prorodeo.com

Research Defense Society
58 Great Marlborough St.
London, W1F 7JY
United Kingdom
www.rds-online.org.uk

Sled Dog Action Coalition
P.O. Box 562061
Miami, FL. 33256
www.helpsleddogs.org

Appendix C

Survival International
11-15 Emerald St.
London WC1N 3QL
England
www.survival-international.org

Texas Snow Monkey Sanctuary
P.O. Box 702
Dilley, TX. 78017

United Animal Nations
5892A South Land Park Drive
P.O. Box 188890
Sacramento, CA. 95818
www.uan.org

Vegan Outreach
211 Indian Dr.
Pittsburgh, PA. 15238
www.veganoutreach.org

Voice for a Viable Future
11288 Ventura Blvd. #202 A
Studio City, CA. 91604
madcowboy.com

VOICE for Animals
P.O. Box 120095
San Antonio, TX. 78212
www.connecti.com/~voice/voice.html

World Council of Whalers
P.O. Box 291
Brentwood Bay, British Columbia
V8M 1R3 Canada
www.worldcouncilofwhalers.com

World Society for the Protection of Animals
P.O. Box 190
Jamaica Plain, MA. 02130
www.wspa.org.uk/home.html

World Wildlife Fund
1250 Twenty-Fourth St. NW
P.O. Box 97180
Washington D.C. 20037
www.worldwildlife.org

Bibliography of Contributors' Writings and Other Resources

Achor, Amy Blount. *Animal Rights: A Beginner's Guide: A Handbook of Issues, Organizations, Actions and Resources*, 2d revised ed. Yellow Springs, OH: Writeware, 452 pp., 1996. (0963186515)

Adams, Carol J. *Neither Man nor Beast: Feminism and the Defense of Animals*. New York: Continuum, 272 pp., 1995. (0826408036)

Adams, Carol J. *The Sexual Politics of Meat: A Feminist-Vegetarian Critical Theory*. New York: Continuum, 256 pp., 1991. (0826405134)

Adams, Carol J., and Josephine Donovan, eds. *Animals and Women: Feminist Theoretical Explorations*. Durham, NC: Duke University, 392 pp., 1995. (0822316676)

Arnold, Ron. *Ecology Wars: Environmentalism as If People Mattered*, reissue edition. Bellevue, WA: Merril Press, 182 pp., 1998. (0939571145)

Arnold, Ron. *Ecoterror: The Violent Agenda to Save Nature: The World of the Unabomber*. Bellevue, WA: Free Enterprise Press, 324 pp., 1997. (0939571188)

Baird, Robert M., and Stuart E. Rosenbaum, eds. *Animal Experimentation: The Moral Issues*. Contemporary Issues series. Buffalo, NY: Prometheus, 182 pp., 1991. (0879756675)

Bauston, Gene. *Battered Birds, Crated Herds: How We Treat the Animals We Eat*. Watkins Glen, NY: Farm Sanctuary, 64 pp., 1996. (0965637700)

Bauston, Gene, and Lori Bauston. *Our Companion Animals: Tales of Transformation from Farm Sanctuary*. New York: Lantern Books, 144 pp., 2001. (1930051239)

Bekoff, Marc. *Strolling with Our Kin: Speaking for and Respecting Voiceless Animals*. New York: Lantern Books, 113 pp., 2000. (1881699021)

Bekoff, Marc, ed. *The Smile of a Dolphin: Remarkable Accounts of Animal Emotions*. Washington, DC: Discovery Channel, 240 pp., 2000. (156331925x)

Bekoff, Marc, and Carron A. Meaney, ed. *Encyclopedia of Animal Rights and Animal Welfare.* Westport, CT: Greenwood, 446 pp., 1998. (0313299773)

Blum, Deborah. *The Monkey Wars.* New York: Oxford University, 306 pp., 1994. (019510109x)

Brebner, Sue, and Debbie Baer. *Becoming an Activist: PETA's Guide to Animal Rights Organizing.* Washington, DC: People for the Ethical Treatment of Animals, 70 pp., 1989.

Bright, Michael. *Intelligence in Animals* (1994). The Earth, Its Wonders, Its Secrets series. New York: Reader's Digest Association Limited, 160 pp., 1997. (0895779137)

Budiansky, Stephen. *The Covenant of the Wild: Why Animals Chose Domestication* (1992). New Haven: Yale University, 212 pp., 1999. (0300079931)

Clutton-Brock, Juliet. *Domesticated Animals from Early Times* (1981), 2d ed. Austin: University of Texas, 208 pp., 1999.

Coats, C. David, and Michael W. Fox. *Old MacDonald's Factory Farm: The Myth of the Traditional Farm and the Shocking Truth about Animal Suffering in Today's Agribusiness.* New York: Continuum, 186 pp., 1989. (0826404391)

Cohen, Robert et al. *Milk: The Deadly Poison.* New York: Argus Archives, 317 pp., 1998. (0965919609)

Crail, Ted. *Apetalk and Whalespeak: The Quest for Interspecies Communication.* Los Angeles: Tarcher, 298 pp., 1981. (0809255278)

Dawne, Diana. *Venture's Story: Life & Times of a Guide Dog.* Orange: Word & Pictures Press, 285 pp., 1997. (0964485737)

DeWaal, Frans. *Good Natured: The Origins of Right and Wrong in Humans and Other Animals*, 3d ed. Cambridge: Harvard University, 296 pp., 1997. (0674356616)

Donovan, Josephine, and Carol J. Adams, ed. *Beyond Animal Rights: A Feminist Caring Ethic for the Treatment of Animals.* New York: Continuum, 216 pp., 1996. (0826408362)

Facklam, Margery. *Wild Animals, Gentle Women.* New York: Harcourt Brace Jovanovich, 139 pp., 1978. (015296987x)

Field, Shelly. *Careers as an Animal Rights Activist.* New York: Rosen, 205 pp., 1993. (0823914658)

Fox, Michael W. *Beyond Evolution: The Genetically Altered Future of Plants, Animals, the Earth . . . and Humans.* New York: Lyons, 256 pp., 1999. (1558219013

Fox, Michael W. *The Boundless Circle: Caring for Creatures and Creation.* Wheaton, IL: Quest, 300 pp., 1996. (0835607259)

Fox, Michael W. *Eating with Conscience: The Bioethics of Food.* Troutdale, OR: New Sage Press, 224 pp., 1997. (0930165309)

Fox, Michael W. *Superpigs and Wondercorn: The Brave New World of Biotechnology and Where It All May Lead.* New York: Lyons & Burford, 209 pp., 1992. (1558211829)

Fox, Michael W., and Cleveland Amory. *Inhumane Society: The American Way of Exploiting Animals* (1990). New York: St. Martins, 268 pp., 1992. (0312042744)

Fraser, Laura, and Joshua Horwitz. *The Animal Rights Handbook: Everyday Ways to Save Animal Lives*. Los Angeles: Living Planet Press, 113 pp., 1990. (0962607207)

Freeman, Milton M.R. *Endangered Peoples of the Arctic: Struggles to Survive and Thrive*. Endangered Peoples of the World series. Westport, CT: Greenwood Publishing Group, 304 pp., 2000. (0313306494)

Freeman, Milton M.R. *Inuit, Whaling, and Sustainability*. Native American Communities paper series. Altamira Press, 200 pp., 1998. (0761990631)

Freeman, Milton M.R., and Urs P. Kreutner. *Elephants and Whales: Resources for Whom?* Gordon & Breach, 1995. (2884490108)

Garner, Robert, ed. *Animal Rights: The Changing Debate*. Politics, Medical Science, Philosophy series. New York: New York Univ., 256 pp., 1997. (0814730981)

Goodall, Jane. *Through a Window: My Thirty Years with the Chimpanzees of Gombe*. Boston: Houghton Mifflin, 268 pp., 1990. (0395599253)

Groves, Julian McAllister. *Hearts and Minds: The Controversy over Laboratory Animals*. Animals, Culture, and Society series. Philadelphia: Temple University, 226 pp., 1997. (1566394767).

Harnack, Andrew, ed. *Animal Rights: Opposing Viewpoints*, reprint ed. San Diego, CA: Greenhaven, 240 pp., 1996. (1565103998)

Hepner, Lisa Ann. *Animals in Education: The Facts, Issues, and Implications*. Albuquerque, NM: Richmond Publishing, 305 pp., 1994. (0963941801)

Hogan, Linda, ed. *Intimate Nature: The Bond between Women and Animals*. New York: Fawcett Columbine, 480 pp., 1998. (0449003000)

Houston, Pam, ed. *Women on Hunting*. Hopewell, NJ: Ecco, 336 pp., 1995. (0880014431)

Hyland, J.R. *God's Covenant with Animals: A Biblical Basis for the Humane Treatment of All Creatures*. Lantern Books, 107 pp., 2000. (1930051158)

Kalechofsky, Roberta. *Autobiography of a Revolutionary: Essays on Animal and Human Rights*. Marblehead, MA: Micah, 189 pp., 1991. (091628834x)

Kalechofsky, Roberta, ed. *Judaism and Animal Rights: Classical and Contemporary Responses* (1992), reprint ed. Marblehead, MA: Micah Publications, 356 pp., 1994. (0916288358)

Kalechofsky, Roberta, ed. *Rabbis and Vegetarianism: An Evolving Tradition*. Marblehead, MA: Micah Publications, 104 pp., 1995. (0916288358)

Kalechofsky, Roberta, ed. *Vegetarian Judaism: A Guide for Everyone*. Marblehead, MA: Micah Publications, 246 pp., 1998. (0916289455)

Marquardt, Kathleen. *Animal Scam: The Beastly Abuse of Human Rights*. Washington, DC: Regnery Gateway, 221 pp., 1993. (0895264986)

Martin, Ann N., and Michael Fox. *Food Pets Die For: Shocking Facts about Pet Food*. Troutdale, OR: New Sage Press, 148 pp., 199. (0939165317)

Mason, Jim, and Peter Singer. *Animal Factories: The Mass Production of Animals for Food and How it Affects the Lives of Consumers, Farmers and Animals*, revised ed. New York: Crown, 174 pp., 1990. (051753844x)

Masson, Jeffrey Moussaieff, and Susan McCarthy. *When Elephants Weep: The Emotional Lives of Animals*. New York: Delacorte, 291 pp., 1995. (0385314256)

McCoy, J.J. *Animals in Research: Issues and Conflicts*. An Impact Book. New York: Franklin Watts, 128 pp., 1993. (0531130231)

Miller, Pat. *Positive Dog Training: The Fun and Rewarding Way to Train Your Dog*. New York: Howell, 230 pp., 2001. (0764536095)

Moretti, Laura. *The Good Fight: Speaking for Those Who Can't*. The Best of the Animals' Voice Magazine series. Chico, CA: MBK Publishing, 82 pp., 1994. (1884873243)

Moretti, Laura A., ed. *All Heaven in a Rage: Essays on the Eating of Animals*. Chico, CA: MBK Publishing, 80 pp., 1999. (1884873146)

Mortenson, F. Joseph. *Whale Songs & Wasp Maps: The Mystery of Animal Thinking*. New York: Dutton, 178 pp., 1987. (0525244425)

Newkirk, Ingrid. *Free the Animals: The Untold Story of the Animal Liberation Front and Its Founder "Valerie."* New York: Noble, 372 pp., 1992. (187936011x)

Newkirk, Ingrid. *You Can Save the Animals: 251 Ways to Stop Thoughtless Cruelty*. Rocklin, CA: Prima, 288 pp., 1999. (0761516735)

Partners in Research. *Biomedical Research: Is It Really Necessary?* (video). Ottawa: Partners in Research, 1992.

Pringle, Laurence. *The Animal Rights Controversy*. New York: Harcourt Brace Jovanovich, 112 pp., 1989. (0152035591)

Rollin, Bernard E. *Animal Rights and Human Morality*, revised ed. Buffalo: Prometheus, 248 pp., 1992. (0879757892)

Rollin, Bernard E. *Farm Animal Welfare: Social, Bioethical and Research Issues*. Ames: Iowa State University, 180 pp., 1995. (0813825636)

Rollin, Bernard E. *The Frankenstein Syndrome: Ethical and Social Issues in the Genetic Engineering of Animals*. Cambridge Studies in Philosophy and Public Policy series. Cambridge: Cambridge University, 241 pp., 1995. (0521478073)

Rollin, Bernard E. *The Unheeded Cry: Animal Consciousness, Animal Pain, and Science*, 2d expanded ed. Ames: Iowa State University, 344 pp., 1998. (081382575x)

Rollin, Bernard E., and M. Lynne Kesel, eds. *The Experimental Animal in Biomedical Research: Care, Husbandry, and Well-Being: An Overview by Species*, 2 volumes. Survey of Scientific and Ethical Issues for Investigators series. Boca Raton, FL: CRC Press, 1995. (0849349818 and 0849349826)

Rowan, Andrew N., ed. *Animals and People Sharing the World*. Hanover, NH: University Press of New England, 192 pp., 1988. (0874514495)

Rowan, Andrew N. *Of Mice, Models, and Men: A Critical Evaluation of Animal Research*. Albany: State University of New York, 323 pp., 1984. (0873957768)

Sherry, Clifford. *Endangered Species: A Reference Handbook*. Contemporary World Issues series. Santa Barbara: ABC-CLIO, 269 pp., 1998. (0874368103)

Sherry, Clifford J. *Animal Rights: A Reference Handbook*. Santa Barbara, CA: ABC-CLIO, 214 pp., 1994. (0874367255)

Shorto, Russell et al. *Careers for Animal Lovers*. Brookfield, CT: Millbrook Press, 64 pp., 1992.

Singer, Peter. *Animal Liberation*, 2d ed. New York: New York Review of Books, 320 pp., 1990. (0940322005)

Sperling, Susan. *Animal Liberators: Research and Morality*. Berkeley: University of California, 247 pp., 1988. (0520061985)

Stange, Mary Zeiss. *Woman the Hunter*. Naples: Beacon, 240 pp., 1997. (0807046388)

Strand, Rod, and Patti Strand. *The Hijacking of the Humane Movement: Animal Extremism*. Wilsonville, OR: Doral, 174 pp., 1993. (0944875289)

Tobias, Michael. *Nature's Keepers: Wildlife Poaching in the U.S. and Efforts to Stop It*. New York: Wiley, 304 pp., 1998. (0471157287)

Tobias, Michael. *Voices from the Underground: For the Love of Animals*. Pasadena, CA: New Paradigm Books, 166 pages, 1999. (0932727484)

Tobias, Michael, and Kate Mattelon, ed. *Kinship with the Animals*. Hillsboro, OR: Beyond Words, 337 pp., 1998. (1885223889)

Williams, Jeanne, ed. *Animal Rights and Welfare*. Reference Shelf vol. 63 no. 4. New York: H.W. Wilson, 168 pp., 1991. (0824208153)

Young, Richard Allen, and Carol J. Adams. *Is God a Vegetarian? Christianity, Vegetarianism, and Animal Rights*. Chicago: Open Court, 187 pp., 1998. (0812693930)

Index

Achor, Amy, 319
Action for Life conference, 149
Activist, definition of, 1, 171, 307
Acton, Lord, 60
Adams, Carol, 13–21, 229, 319, 323
Adams, Richard, 139
Adler, Margot, 27
Advances in Animal Welfare Science (magazine), 106
African Primate Specialists Group, 235
"Agenda 21," 183
Agnosticism, 59–60
Ahimsa, 133, 229–30
AIDS, 49, 162
All Creatures Great and Small, 226
All Heaven in a Rage: Essays on the Eating of Animals, 201, 322
All the President's Men (movie), 124
Allevato, Diane, 197
Alliance for America (AFA), 58, 222, 247–48, 311
American Agri-Women, 222
American Association for the Accreditation of Laboratory Animal Care (AAALAC), 286
American Association of Equine Practitioners, 255
American Cancer Society, 148
American Horse Council, 254, 256
American Humane Association (AHA), 256, 311
American Quarter Horse Association, 256
American Society for the Prevention of Cruelty to Animals (ASPCA), 100, 122, 256, 311
American Veterinary Medical Association, 255
Americans for Medical Progress, 311
Amory, Cleveland, 321
Animal-assisted therapy, 92–99. *See also* Guide dogs
Animal Experimentation, 319
Animal experimentation. *See* Vivisection
Animal Factories, 322
Animal Forum (radio program), 227
Animal intelligence. *See* Intelligence of animals
Animal Legal Defense Fund, 299
Animal Liberation, 73, 194, 209, 229, 238–39, 323
Animal Liberation Front (ALF), xi, 7, 23, 29–35, 139, 143, 211, 266. *See also* Direct-action activism; Extremism

Animal Liberation Front Press Office, 312
Animal Liberators, 323
Animal Machines, 200
Animal Protection Institute (API), 38, 50, 122, 312
Animal Rights: A Beginner's Guide, 319
Animal Rights: The Changing Debate, 321
Animal Rights: Opposing Viewpoints, 321
Animal Rights: A Reference Handbook, 323
Animal Rights: A Subject Guide, Bibliography, and Internet Companion, 11, 307
Animal rights, definition of, 2, 5, 140, 284
Animal Rights Advocates of Upstate New York, 131, 311
Animal Rights and Human Morality, 322
Animal Rights and Welfare, 323
The Animal Rights Controversy, 322
Animal Rights Handbook, 321
Animal Rights Legislative Action Network, 215, 312
Animal Rights Network, 229
Animal Rights Online (newsletter), 2, 228–29, 312
Animal sanctuaries, 20, 36, 124
Animal Scam, 322
Animal welfare, definition of, 2–5, 140, 284
Animal Welfare Institute, 299, 312
Animal's Agenda (magazine), 17, 50, 202, 229–30, 312
Animals and People Sharing the World, 323
Animals and Society (radio show), 227
Animals and Women, 319
Animals as entertainment. *See* Circuses; Rodeos
Animals in Education, 321
Animals in Research, 322
The Animals' Voice (magazine), 201–02
Animals' Voice Online, 201, 312
Animism, 110
Anthony, Susan B., 18
Anthropocentrism, 43, 50, 74, 107, 234, 296
Apetalk and Whalespeak, 320
Applied Research Ethics National Association, 286
Aquariums. *See* Zoos
Aristotle, 26, 267, 271
Ark Trust, 122, 215, 312
Arnold, Ron, 22–28, 319
Association for Laboratory Animal Science, 285
Association of Pet Dog Trainers, 195
Association of Veterinarians for Animal Rights (AVAR), 122, 312
Atheism, 38
Autobiography of a Revolutionary, 160–61, 321

Bach, Richard, 134
Baer, Debbie, 320
Baird, Robert, 319
Barbarash, David, 29–35
Bardot, Brigitte, 137
Barnard, Neal, 62, 209
Barnes, Don, 36–40, 208
Barry, Wendell, 297
Battered Birds, Crated Herds, 319
Bauston, Gene, 41–44, 276, 319
Bauston, Lori, 319
Beautiful Joe, 194
Becoming an Activist, 320
Bekoff, Marc, 45–51, 276, 319–20
Bentham, Jeremy, 270
Berry, Rynn, 229–30
Berry, Thomas, 109, 185
Berry, Wendell, 109
Between the Species (journal), 161
Beyond Animal Rights, 320

Beyond Evolution, 320
Bias, 6, 57, 80, 113, 115, 171
Bible, the, 75, 84, 90, 154–57, 164, 185, 204, 210, 230, 248
Biodiversity, 183
Biomedical Research, 322
The Biosynergy Institute, 235
Bishop, Brian, 52–61
Bishop of Manchester, 85
Black Beauty, 194
Black Cat Video, 30
Black Elk, 109
Blum, Deborah, 320
The Boundless Circle, 110, 320
Bovine growth hormone, 62, 64
Boys Town National Research Hospital, 289, 312
Brebner, Sue, 320
Bright, Michael, 320
Brophy, Brigid, 238
Brower, David, 26
Brundtland, Gro Harlem, 185
Buddhism, 97–98, 117, 270, 275
Budiansky, Stephen, 320
Burke, Edmund, 79
The Bushmeat Crisis Taskforce, 235
The Bushmeat Project, 235, 312
Butterfly, Julia, 299

California Animal Control Directors' Association, 196
Canadian Royal Commission on Sealing, 141
Careers as an Animal Rights Activist, 320
Careers for Animal Lovers, 323
Carson, Rachel, 270
The Case for Animal Rights, 2, 140, 190
Cats, 47–48, 93, 104
Center for Animals and Public Policy, 237–38
Center for the Defense of Free Enterprise, 22–25, 313
Cetacean Freedom Network, 299

CHAIN Letter (newsletter), 195
Chattanooga Humane Educational Society, 196
Chickens. *See* Poultry
Chief Seattle, 185
Christianity and the Rights of Animals, 230
Church of the Earth, 300
Churchill, Winston, 277
Circuses, 32, 37, 101, 120, 123, 128, 162
Citizens for Responsible Animal Behavior Studies, 51
Classen, Tom, 120
Cloning of animals, 110. *See also* Genetic engineering
Clothing from animals, 32, 128, 138–39, 150, 156, 162, 188, 219–25, 274. *See also* Fur farming
Clutton-Brock, Juliet, 320
Coalition to Abolish the Fur Trade, 169, 267, 313
Coats, David, 320
Cock-fighting, 102
Cohen, Robert, 62–71, 320
Cohn, Priscilla, 72–77
Collective Humane Action and Information Network, 195, 313
Communal Areas Management Programme for Indigenous Resources, 56
Communication, between animals and humans. *See* Interspecies communication
Compassion for Animals Foundation, 202, 313
Comstock, Rose, 223
Conservation. *See* Sustainable use
Conservation International, 274
Convention on International Trade in Endangered Species (CITES), 299
Coronado, Rodney, 23, 97, 299
Cousins, Norman, 86

The Covenant of the Wild, 320
Coyne, Karen, 78–86
Crail, Ted, 320
Cranford, Mike, 121
Crazy Woman Bison Ranch, 259
Crimes, against animal enterprises. *See* Animal Liberation Front (ALF); Direct-action activism; Extremism
Curtis, Stanley, 110

Da Vinci, Leonardo, 132, 276
Dairy Education Board, 66, 313
Dairy products, 62–71, 126, 168, 173, 244, 246
Dalai Lama, 142
Darwin, Charles, 47, 75
Davis, Gail, 68
Davis, Karen, 20–21
Dawne, Diana, 87–91, 320
de Chardin, Teilhard, 109, 185
DeBakey, Michael, 281
Decker, Carlene, 286
A Declaration of War, 23
"Deep Ethology," 49
DeMares, Ryan, 92–99
The Descent of Man, 47
DeVillars, John, 60
DeWaal, Frans, 320
Direct-action activism, 23–35, 39, 57, 86, 97, 139, 147, 150, 158, 188, 191, 210–11, 214, 222–23, 225, 247, 255, 266–27, 282, 287, 292–93, 300–302. *See also* Animal Liberation Front (ALF); Extremism
Dissection. *See* Vivisection
Dog sledding, 119–25
Dogs, 47–48, 100, 104, 119–25, 127–28, 174–75, 193–200, 265–66, 287
Dollase, Vern, 84–85
Dolphin Alliance, 313
Dolphin Connection, 37
Dolphin Embassy USA, 37

Dolphin Institute, 97
Dolphins, 37, 54–55, 92–99, 219, 295–303
Domesticated Animals from Early Times, 320
Dominion/domination (as religious concept), 19, 38, 97, 154, 170
Dommer, Luke, 76
Donovan, Josephine, 319–20
Doris Day Animal League, 122, 313
Dostoyevski, Fyodor, 185
Doumar, Robert, 250–51
Drawing Down the Moon, 27
Durbin, Sherrill, 100–103

Earth First! 24, 191, 299–300. *See also* Direct-action activism
Earth Island Institute, 26
Earth Liberation Front, 32. *See also* Direct-action activism
EarthSave, 215
Eating with Conscience, 320
Ecofeminism and the Sacred, 19
Ecofeminism. *See* Feminism
Ecology Wars, 319
Ecoterror, 24–25, 319
EcoTerror Response Network, 24
Eggs. *See* Poultry
Ehrlich, Anne, 276
Ehrlich, Paul, 276
Einstein, Albert, 203
Eisnitz, Gail, 62–63
Elephants and elephant hunting, 56, 109, 175, 179
Elephants and Whales, 321
Eliot, T.S., 164
The Emporer's New Clothes, 154
The Encyclopedia of Animal Rights and Animal Welfare, 51, 320
Endangered Peoples of the Arctic, 321
Endangered Species: A Reference Handbook, 323
Endangered Species Act, 251
Environmental Protection Agency (EPA), 58, 60

Environmentalism, related to animal issues, 32, 53–55, 80, 118, 146, 161, 189, 201, 220, 224, 227, 234, 260, 272, 300
The Epic of Gilgamesh, 25
Ethologists for the Ethical Treatment of Animals, 51
European Bureau for Conservation and Development (EBCD), 184
The Experimental Animal in Biomedical Research, 323
Extinction of species, 108, 137, 189, 270, 272
Extremism, 1, 7, 23–24, 30, 171, 188, 266. See also Direct-action activism

Facklam, Margery, 320
Factory farming, 42, 64–65, 79, 108–10, 129, 140–41, 159, 161, 194–95, 201–02, 227–28, 259–60, 271
Farm Animal Reform Movement (FARM), 149, 313
Farm Animal Welfare, 322
Farm Sanctuary, 42, 65, 67, 81, 85, 130, 274, 313
Farming, of animals. See Factory farming
Fatal Exposure, 273
Federation of American Societies for Experimental Biology, 241, 314
Feminism, as related to animal issues, xii,13–21, 258–64
Feminists for Animal Rights, 17–18, 262–63, 314
Field, Shelly, 320
Fishermen's Coalition, 219, 314
Fishing and fisheries, 54–55, 178, 219, 221, 224, 227–28, 244–52, 272, 291
Flexner, Simon, 49
Florida Voice for Animals, 227
Food and Drug Administration (FDA), 62, 65–66
Food for the Gods, 229–30
Food Pets Die For, 322

Foot and Mouth disease, 174, 271
Ford, Rich, 286
Foundation for Biomedical Research (FBR), 280–81
Fox, Michael, 104–11, 208, 320–22
Francis of Assisi, St., 76, 132, 197, 276, 299
Frank, Morris, 89
The Frankenstein Syndrome, 322
Fraser, Laura, 321
Fred Coleman Memorial Labor Day Pigeon Shoot, 130
Free the Animals!, 211, 322
Freeman, Milton M.R., 112–18, 321
Friends of Animals, 262, 314
Friends of Rodeo Newsletter, 255
Friends of the Earth, 26
Frontline Information Service, 32
Fund for Animals, 75, 156, 202, 260, 267, 314
Fund for the Replacement of Animals in Medical Experiments, (FRAME), 237–38, 314
Fur: The Fabric of a Nation (video), 141
Fur Commission USA, 169, 220, 314
Fur Council of Canada, 141, 314
Fur farming, 31–32, 35, 135–44, 156, 168, 219–25. See also Clothing from animals

Gandhi, Mahatma, 43, 81, 83, 117, 149, 216, 277, 294, 299
Garner, Robert, 321
Genetic engineering, 62–63, 68, 105–06, 110. See also Cloning of animals
Genetically modified foods, 29, 32, 62
Genetix Alert Press Office, 32
Gibran, Kahlil, 89
Glacken, Clarence, 271
Glickman, Margery, 119–25
Global Communications for Conservation, 177
God's Covenant with Animals, 321

The Good Fight, 322
Good Natured, 320
Goodall, Jane, 50, 321
Gorall, Kimber, 126–34
Gore, Albert, 60, 124
The Gorilla Foundation, 235
Gottlieb, Alan, 25
Great American Meatout, 147
Greenpeace, 3, 137, 185, 221, 276, 314
Groves, Julian, 321
GRRR! (newsletter), 123
Guide dogs, 87–91. *See also* Animal-assisted therapy
Gun Women, 261

Habermas, Jurgen, 26
Haggadah for the Liberated Lamb, 164
Haggadah for the Vegetarian Family, 164
Han, Tich Nhat, 43, 299
Harnack, Andrew, 321
Harrison, Ruth, 200
Health and Healing (newsletter), 62
Hearts and Minds, 321
Hegarty, Dorothy, 237
Heimlich, Jane, 62
Hepner, Lisa, 321
Herbert, Agnes, 263
Herriot, James, 226
Herscovici, Alan, 135–44
Hershaft, Alex, 145–51
The Hijacking of the Humane Movement, 323
Hill View Farm Animal Refuge, 174, 176, 180–81
Hinduism, 117
Hippocrates, 216
Hitler, Adolf, 162, 277
Hoard's Dairyman (magazine), 64
Hogan, Linda, 321
Horwitz, Joshua, 321
Houston, Pam, 321
Human-animal communication. *See* Interspecies communication

Humane Farming Association, 195, 314
Humane Religion, 155
Humane Slaughter Act, 147. *See also* Laws, for animals
Humane Society of the United States (HSUS), 75, 99, 104–11, 122, 142, 237–43, 256, 267, 293, 314
Humane Society University, 239
Hunting, 32, 73–74, 128, 130, 133, 154–55, 170, 178, 182–87, 208, 258–64, 290–91
Hyland, J.R., 152–58, 321

Iditarod, dog sled race, 119–25
Iditarod Trail Committee, 122, 315
In Defense of Animals, 36, 190
In Defense of Animals (IDA), 101, 122, 202, 256, 315
India Project for Animals and Nature (IPAN), 105, 175–81, 315
Indigenous peoples, rights of, 108, 112–18, 135–44, 182–86, 239, 298
Indigenous Survival International, 184
Inhumane Society, 321
The Inner Art of Vegetarianism, 14, 16, 19, 229
Institute for the Study of Animal Problems, 106, 237
Institutional Animal Care and Use Committee (IACUC), 284
Intelligence in Animals, 320
Intelligence of animals, 39, 46–47, 67, 73, 76–77, 85, 88, 90, 94, 96, 100–101, 128, 168, 189, 272
Intensive animal agriculture. *See* Factory farming
International Fund for Animal Welfare (IFAW), 137, 142, 184–85, 315
International Journal for the Study of Animal Problems (magazine), 106
International Union for the Conservation of Nature (IUCN), 116, 184, 315

Index

International Whaling Commission, (IWC) 55–56, 113, 184, 297–99, 315
International Wildlife Management Consortium (IWMC), 116
International Working Group for Indigenous Affairs (IWGIA), 184
Interspecies communication, 39, 92–99
Intimate Nature, 321
Inuit Circumpolar Conference, 116, 184, 315
Inuit, Whaling, and Sustainability, 321
Is God a Vegetarian?, 323
Islam, 117, 155, 275

Jainism, 275–76
Jane Goodall Institute, 50
Jefferson, Thomas, 26
Jesus. *See* Bible, the
Jews for Animal Rights, 160, 315
Joan of Arc, 299
Journal of Applied Animal Welfare Science, 195
Judaism, 34, 117, 153, 155, 159–65, 275–76
Judaism and Animal Rights, 321
Judaism and Vegetarianism, 159

Kalechofsky, Roberta, 159–65, 321
Kendall, Crystall, 166–71
Kesel, Lynne, 323
King, Angus, 60
King, Martin Luther, Jr., 43, 117, 267
Kinship with the Animals, 323
Kistler, John M., xi, 107, 322
Knight, Les U. (pseudonym), 189
Kohlberg, Lawrence, 82
Koran, the, 84
Krantz, Deanna, 105, 172–81
Kreutner, Urs, 321
Kyi, Aung San Suu, 299

Last Chance for Animals, 122, 202, 315
LaVey, Anton, 197–98

Laws, for animals, xi, 56, 73, 86, 101, 107, 120–21, 123, 130–31, 176, 208, 246, 251, 254–55, 273–74, 281, 285–86. *See also* Endangered Species Act; Humane Slaughter Act; *specific laws*
Leary, Timothy, 136
Leather. *See* Clothing from animals; Fur farming
Leffingwell, Albert, 241
Licenses for Animal Welfare, 196
Lincoln, Abraham, 216, 267, 288, 294
Linzey, Andrew, 230
Living Among Meat Eaters, 16
Low Frequency Active Sonar (LFAS), 296–98
Lucretius, 271
Lyman, Howard, 62, 67, 132, 149, 230
Lynge, Finn, 182–86

McCartney, Linda, 143
McCartney, Paul, 143
McCoy, J.J., 322
Mad Cow disease, 148, 230, 274
Mahavira, 276
Malaparte, Curzio, 159–60
Mandela, Nelson, 43, 185
Mara, Jack, 286
March for Animals, 130
Marin Humane Society, 197
Marine Mammal Protection Act, 142
Marquardt, Kathleen, 187–92, 322
Martin, Ann, 322
M*A*S*H* (television program), 179, 225
Maslow, Abraham, 95
Mason, Jim, 322
Masson, Jeffrey, 322
Mattelon, Kate, 323
Matthiessen, Peter, 270, 276
Mead, Margaret, 86
Meaney, Carron, 320
Meat production, 15, 53, 65, 70–71, 78, 129, 132, 145–46, 174, 196, 200–202, 227–28, 246, 271, 274

Media, 4–6, 29–30, 39, 79–81, 113, 115, 118, 124, 137, 139–41, 183, 201, 224, 240, 242–43, 261
Medical Research Modernization Committee (MRMC), 49–50, 315
Medina, Harold, 223
Milk. *See* Dairy products
Milk: The Deadly Poison, 62, 320
Mill, John Stuart, 83
Miller, Pat, 193–98, 322
Miller, Paul, 197
Minding Animals, 51
The Monkey Wars, 320
Monkeys. *See* Primates
Moore, Vicky, 76
Moretti, Laura, 199–205, 322
Morrison, Jane Gray, 276
Mortenson, F. Joseph, 322
Muir, John, 276
Mushing, 119–25

Nansen, Fridtjof, 117
National Animal Interest Alliance (NAIA), 142, 222, 254, 256, 265–68, 293, 316
National Anti-Vivisection Society (NAVS), 316
National Association for Biomedical Research, 241, 280–81, 293, 316
National Cancer Institute, 148
National Conference of State Legislators, 256
National Council on Pet Population, 195
National Fisheries Institute, 247
National Institutes of Health, 241
National Marine Fisheries Service, 250–51
National Veal Ban Action, 149
Nature and Peoples of the North, 184
Nature's Keepers, 323
Navarre, Patrick, 286
Neither Man nor Beast, 319
Newkirk, Ingrid, 23, 206–11, 216, 268, 274, 322

Nietzche, xii
No Compromise (magazine), 32
North American Vegetarian Society, 229
North American Veterinary Technician Association (NAVTA), 285–86
North Carolina Agribusiness Council, 247
North Carolina Fisheries Association, 246, 316

Oakley, Annie, 258, 263
Of Mice, Models, and Men, 323
Office of Laboratory Animal Welfare, 285–86
Old MacDonald's Factory Farm, 320
One, 134
Orange County People for Animals, 213–16, 316
Otter, Nigel, 180–81
Overpopulation of animals, 37, 118, 175–76, 195–96, 284
Overpopulation of humans, 38, 272–73
Oyster, Carol, 261

Paganism, 34
Pain, in animals, 49, 67–68, 77, 101–02, 106–7, 109, 116, 195, 227, 238–40, 271, 296
Pantheism/Panentheism, 34, 68, 110, 179, 197, 203
Park, Ava, 212–18
Partners in Research, 322
Peaceable Paws, 193
People for the Ethical Treatment of Animals (PETA), 67, 75, 87, 101, 122–23, 140–42, 156, 185, 187, 206–11, 215, 247–48, 256, 260, 263, 267, 274, 293, 299, 316
Pet shops, 100, 106–7, 120, 127, 197, 265, 284
Physicians Committee for Responsible Medicine, 62, 67, 209, 316

Pigs, 39, 53, 110, 169, 200–202, 206
Pikes Peak Humane Society, 256
Pipkorn, Carol, 223
Pipkorn, Tom, 223
Pity Not Cruelty (PNC), 75
Platt, Teresa, 219–25
Plutarch, 161, 216
Politically Correct Environment, 27
Positive Dog Training, 193, 322
Poultry, 21, 41, 68, 107, 112, 210, 244
Powys, John Cowper, 161
Primates, 23, 36, 38–39, 49, 96, 232–36
Pringle, Laurence, 322
Pro Rodeo Sports News, 255
Professional Rodeo Cowboys Association (PRCA), 253–56, 316
The Prophet, 89
Psychological well being of activists, 10, 93–94, 102, 134, 170, 179–80, 198, 215, 217–18, 248, 264, 297, 302
Publicity. *See* Media
Puppy mills. *See* Pet shops
Pythagoras, 132, 294

Quixote, Don, 38, 55, 59, 62

Rabbis and Vegetarianism, 321
Raffi, 134
Rand, Ayn, 59
Ray, Dixie Lee, 191
Regan, Tom, 2, 140, 190
Religion. *See specific religions*
Research Defense Society, 242, 316
Rights, concept of, 27–28, 56–59, 113–14, 118, 190–91
Rio Declaration, 183
Robbins, John, 79, 229
Robbins, Tom, 38
Rochester Area Vegetarian Society Inc., 132
Rocky Mountain Animal Defense, 50
Rodeos, 32, 37, 129, 253–57

Roghair, Susan, 226–31
Rollin, Bernard, xi–ii, 46, 322–23
Rose, Anthony, 232–36
Rosenbaum, Stuart, 319
Rowan, Andrew, 106, 237–43, 322–23
Russell, Bertrand, 117
Russian Association for the Indigenous Peoples of the North (RAIPON), 184

Sacrifice of animals (religious), 153–54, 197–98
Salt, Henry, 241
Santeria, 197
Saraceno, Jon, 120
Satanism, 161, 197–98
Schill, Jerry, 244–52
Schliefer, Harriet, 161
Schonholtz, Cindy, 253–57
Schramm, Mikayla, 123–24
Schwartz, Richard, 159–60
Schweitzer, Albert, 76, 117, 197, 216, 269
Screaming Wolf (pseudonym), 23
The Sea Shepherd Society, 267
Seal hunting, 56, 112–16, 137–38, 182–83, 186, 200–201
Second Nature, 137–38, 141, 143
The Sexual Politics of Meat, 13–14, 16, 19–20, 319
Shaftesbury, Lord, 157
Shaw, George Bernard, 84, 86
Sherry, Clifford, 323
Shorto, Russell, 323
Sierra Club, 26
Silver Springs monkeys case, 23
Singer, Peter, 36, 73, 190, 194, 209, 229, 238–39, 322–23
The Sixth Day of Creation, 160
Skin, 159–60
The Sky's On Fire (movie), 273
The Slaughter of the Terrified Beasts, 154
Slaughterhouse, 62–63
Slaughterhouses. *See* Meat production

Sled Dog Action Coalition, 122, 316
Sled dogs, 119–25
The Smile of a Dolphin, 319
Society of Animal Welfare Administrators, 196
Socrates, 86, 111
Southern California Primate Research Forum, 235
Soy products, 69
Speciesism, 33, 37, 43, 49, 99
Sperling, Susan, 323
Spira, Henry, 21, 50, 149, 241
Spock, Dr., 64
Stange, Mary Zeiss, 258–64, 323
Stanton, Elizabeth Cady, 18
Stram, Veda, 215
Strand, Patti, 265–68, 323
Strand, Rod, 323
Strolling With Our Kin, 51, 319
Superpigs and Wondercorn, 320–21
Survival International, 317
Sustainable use, 26, 53–55, 58, 107, 115–16, 118, 185, 219–25, 299
Sweeney, Jay, 302

Temple, Lorraine, 121
The Ten Trusts, 51
Tennessee Association of Positive Pet Trainers, 196
Terrorism. *See* Direct-action activism; Extremism
Texas Snow Monkey Sanctuary, 36–39, 317
Thales, 271
Thompsen, Frederick, 241
Thoreau, Henry David, 86, 216
Through a Window, 321
Tobias, Michael, 269–78, 323
Tolstoy, L, 70–71, 109
Traces of the Rhodian Shore, 271
Trapping, 32, 37, 56, 75, 114, 138–39, 208, 239. *See also* Clothing from animals; Fur farming
Trull, Frankie, 279–82
Truth, Sojourner, 208

Tufts University School of Veterinary Medicine, 237
Tzu, Lao, 109, 276, 299

UCLA Brain Research Institute, 233
Unabomber, 25
UNESCO, 141, 176
The Unheeded Cry, 322
Union of Concerned Scientists, 161
United Animal Nations, 101, 122–23, 317
United Nations Environment Program, 139
United Poultry Concerns, 20–21
Utah Animal Rights Coalition, 169
Utilitarianism, philosophy, 47–48
Utley, Laura, 177

Veal. *See* Meat production
Vegan Outreach, 38, 317
Veganism, 14, 17, 37, 42, 63, 69, 78–86, 128–30, 148, 150, 166–71, 176, 230
Vegetarian Judaism, 321
Vegetarian Society of Tampa Bay, 227
Vegetarianism, 2, 13–21, 25, 63, 69, 79, 130, 148, 156, 162, 166–71, 176, 200–201, 204, 206, 210, 228–30, 262, 274, 277
Venture's Story, 87, 89, 320
Veterinarians, 37, 76, 105–08, 123, 176, 193, 271, 279, 283–87
Vincent, Bruce, 223
Violence, relationship between human and animal, 13, 31, 94, 155, 185, 214
Violence and activism. *See* Direct-action Activism; Extremism
Visionaries (radio show), 214
Vivisection, 32, 35, 47–50, 63, 70, 74, 94, 104, 111, 129, 152–53, 159–63, 178, 208, 211, 213, 228, 233–34, 237–43, 279–94
Voice for a Viable Future, 67, 317
VOICE for Animals, 38, 317

Voices from the Underground, 276, 323
Voltaire, 216
Voluntary Extinction Movement, 189

Wade, William, 283–87
Wagner, Thomas, 110
Waiting at the Edge (video), 142
Wallace, Brad, 51
Wallace, William, 299
Walsh, Ed, 288–94
Walsh, John, 203
Watership Down, 139
Western Horseman Magazine, 255
Whale Songs and Wasp Maps, 322
Whales and whaling, 3, 55–56, 107, 112–16, 183, 186, 295–303
What Are People For?, 297
When Elephants Weep, 322
White, Ben, 295–303
Whole Dog Journal, 193
Wicca, 34
Wiesel, Elie, 134
Wilberforce, William, 157
Wild Animals, Gentle Women, 320
Wildlife, 46, 48, 74, 92–99, 116, 118, 132, 259–62, 269–78

Wildlife Poaching in the U.S. and Efforts to Stop It, 323
Wildlife Protectors Fund, 235
Williams, Jeanne, 323
Wilson, Edward O., 59, 276, 293–94
Wise, Stephen, 275
Wise use. *See* Sustainable use
Woman the Hunter, 259–60, 323
Women on Hunting, 321
Woodward, Bob, 124
World Conservation Strategy, 139
World Council of Whalers, 116, 317
World Farm Animals Day, 149
World Society for the Protection of Animals, 203, 317
World Trade Organization (WTO), 108, 297
World Vegetarian Congress, 146
World Wide Fund for Nature, 184
World Wildlife Fund, 92, 276, 299, 317
Wright, Phyllis, 197

You Can Save the Animals, 210–11, 322
Young, Richard Allen, 323

Zoos, 32, 37, 48, 50, 128–29, 162

About the Contributors

CAROL J. ADAMS is the author of *The Sexual Politics of Meat: A Feminist-Vegetarian Critical Theory*, now in a Tenth Anniversary Edition, and of *Living among Meat Eaters: The Vegetarian's Survival Handbook*.

RON ARNOLD is executive vice president of the Center for the Defense of Free Enterprise in Bellevue, Washington, and is the author of seven books on environmental topics.

DAVID BARBARASH is an environmental and animal liberation activist. He is a former member of the underground Animal Liberation Front, convicted of rescuing cats from a university research facility, and now acts as its public spokesperson.

DON BARNES, a former animal researcher, is the Southern Field Representative for the Animal Protection Institute and executive director of VOICE for Animals, Inc.; Don is a twenty-year veteran of the animal rights movement.

GENE BAUSTON is cofounder and director of Farm Sanctuary, a farm animal protection organization. He visits farms, stockyards, and slaughterhouses to document and expose animal abuse. He works to promote new laws and to utilize existing laws to protect farm animals.

MARC BEKOFF is professor of Biology at the University of Colorado, Boulder. He and Jane Goodall have cofounded the organization Ethologists for the Ethical Treatment of Animals: Citizens for Responsible Animal Behavior Studies (*www.ethologicalthics.org*).

BRIAN BISHOP works with Rhode Island Wiseuse to promote sustainable use in public environmental plans sponsored by the government.

ROBERT COHEN is the executive director of the Dairy Education Board. He has a degree in Psychobiology, and once performed live animal research in the field of neuroendocrinology. He lectures about the dangers of milk and dairy products, and believes that animal research is a betrayal to both humans and animals.

PRISCILLA COHN, Ph.D., professor of Philosophy at Abington College, the Pennsylvania State University, has written and lectured on five continents about moral issues concerning nonhuman animals.

KAREN COYNE promotes a vegan lifestyle and opposes factory farming.

DIANA DAWNE is a blind person who writes and speaks about the importance of guide dogs for seeing-impaired people.

RYAN DEMARES is a writer, interspecies workshop leader, and a director of The Dolphin Institute, a Washington State nonprofit organization.

SHERRILL DURBIN is a native Oklahoman who currently works as a purchasing agent for a community college, spends her spare time doing animal rights activities, and shares her home with five rescued dogs.

MICHAEL FOX is a veterinarian, ethologist, syndicated columnist, and is senior scholar for bioethics with The Humane Society of the United States.

MILTON M.R. FREEMAN is an ecologist with an interest in the sustainable and equitable use of natural resources.

MARGERY GLICKMAN is a retired elementary schoolteacher who founded the Sled Dog Action Coalition, a group that educates the public about the cruelties of the Iditarod dog sled race.

KIMBER GORALL is a writer, activist, and president of Animal Rights Advocates of Upstate New York, Inc.

ALAN HERSCOVICI has researched and written about environmental and human rights issues for more than twenty years; he is now Executive Vice-President of the Fur Council of Canada.

ALEX HERSHAFT, Ph.D., is the founder and president of FARM (Farm Animal Reform Movement), founder and chair of the Great American Meatout, a cofounder of the U.S. animal rights movement, and chair of the movement's annual conference.

About the Contributors

J.R. HYLAND is an Evangelical minister, theologian/author, and director of Humane Religion, an educational and outreach ministry.

ROBERTA KALECHOFSKY is an accomplished fiction writer whose work has been translated into Italian, as well as a publisher and animal rights activist.

CRYSTAL KENDELL is currently a student of geology at Utah State University in Logan, Utah. She is an active member of the Utah Animal Rights Coalition and president of the college club "the Student Animal Liberation Team."

JOHN M. KISTLER is Acquisitions Librarian at Utah State University and the author of *Animal Rights: A Subject Guide, Bibliography, and Internet Companion.*

DEANNA KRANTZ is founder and director of India Project for Animals and Nature, a division of Global Communications for Conservation, Inc., New York (*www.gcci.org*).

FINN LYNGE supports biologically sustainable harvesting of marine mammals out of respect for cultural diversity.

KATHLEEN MARQUARDT is a writer who focuses on the radical animal rights movement, its promotion of animal rights over animal welfare, and its attack on humans who use animals through medical research or as pets.

PAT MILLER is a positive pet trainer, behavior consultant and writer whose mission includes promoting a philosophy of respect for life, owner of Peaceable Paws Dog & Puppy Training in Chattanooga, Tennessee, and a retired animal protection professional (Marin Humane Society, 1976–1996).

LAURA MORETTI is a writer and graphics designer of animal rights material, and is founder/director of Animals Voice Online.

INGRID NEWKIRK is founder and president of People for the Ethical Treatment of Animals (PETA), and a vigorous proponent of the idea that human beings are but one animal nation and must not treat the others cruelly.

AVA PARK is the founder of Orange County People for Animals, a nationally known Southern California animal advocate organization, and a radio show host and public speaker on the topics of animal rights and the interconnectedness of all life.

TERESA PLATT works with Fur Commission USA to support humane and sustainable use of animals and animal products.

SUSAN ROGHAIR is the president of the Internet's largest animal rights list, Animal Rights Online; on the advisory board of *The Animals' Agenda* magazine; and a board member of her local animal rights group, Florida Voices for Animals.

BERNARD ROLLIN is professor of Philosophy and professor of Physiology at Colorado State University, where he developed the world's first course for veterinarians on veterinary ethics and animal ethics. He has written extensively and lectured all over the world on animal ethics, animal consciousness, animal pain, farm animal welfare, genetic engineering, cloning and other issues in bioethics.

ANTHONY L. ROSE is a social psychologist and conservation educator who is working to restore the biosynergy of humanity and nature.

ANDREW ROWAN is senior vice president of the Humane Society of the United States and a student of the animal protection movement and human-animal interactions.

JERRY SCHILL is the president of the North Carolina Fisheries Association, Inc., a nonprofit trade association representing the interests of its member commercial fishermen, seafood dealers, and processors.

CINDY SCHONHOLTZ is the animal welfare coordinator for the Professional Rodeo Cowboys Association in Colorado Springs, Colorado.

MARY ZEISS STANGE teaches women's studies and religion at Skidmore College, and writes about women, hunting, and feminist and environmental theory.

PATTI STRAND works with purebred dogs, cofounded the National Animal Interest Alliance, and wrote the book *Hijacking of the Humane Movement: Animal Extremism*.

MICHAEL TOBIAS, ecologist and filmmaker, is the author of some twenty-five books, and the writer, director, and producer of more than one hundred films, many focusing upon biodiversity, public policy, and empathy.

FRANKIE L. TRULL works with the Foundation for Biomedical Research to promote humane use of animals in important medical experiments to better the lives of humans and animals.

WILLIAM L. WADE is a licensed veterinary technician and certified laboratory animal technologist; he is currently the manager of compliance and training at the Center for Comparative Medicine at Northwestern University.

ED WALSH is a professor in Otolaryngology at the Creighton University School of Medicine in Omaha, Nebraska; director of the Developmental Auditory Physiology Laboratory at the Boys Town National Research Hospital; and serves on the Board of Directors for the National Animal Interest Alliance.

BEN WHITE, international coordinator of the Animal Welfare Institute and creator of the WTO turtles, is a single father of two who straddles the worlds of mainstream diplomacy and direct action, working mainly in the arenas of whale and dolphin protection and the preservation of ancient forests.